U0054449

異國風主食料理

焗烤、燉飯、粥品、鍋物等
60道美味幸福上桌（中英對照）

作者：洪白陽（CC老師）
攝影：楊志雄

作者序

料理讓人生更加美好

這本書的內容CC非常喜歡；也是CC一直很想寫的，可能是橘子文化聽到了CC內心的呼喚吧！謝謝出版社。哈哈哈，超有默契的！

這本書的料理反映了現在很多家庭或單身者的需求；只要做個主食就可以在餐桌上展現出豐盛美味的料理。

每出一本書都會得到許多讀者對CC的支持和讚賞，常常令CC開心到如飛越雲端，這也讓CC更加努力，更激發出要設計出更多讓大家喜愛的美食。

CC常想，CC是非常幸運的，因為美食而認識了很多愛護我的讀者和學生，因為有你們讓CC在生活中在人生中更加美好；謝謝大家，有你們真好！

目　錄

PART 1 瀰漫異國風味的歐式料理

PART 2 充滿南洋風味的東南亞料理

※ 依教育部重編國語辭典修訂本，「番茄」與「蕃茄」通用，本書內容均以「番茄」為字。

PART 3 令人食指大動的日韓料理

PART 4 中港台料理

CC 老師料理的省時好幫手

想要和 CC 老師一樣優雅愜意的做菜，好用的廚房用具絕對不可少，做起料理事半功倍，美味營養更是滿分！

子母壓力鍋 4L+8L

壓力鍋為廚房必備鍋具之一，適合用來烹調久煮不易熟的食材，非常簡單又省時。8L 的大容量更可一次三道菜，燉肉、煮湯、煮飯一次到位，時間、健康、美味都顧到了！

完美雙享壓力鍋 5L

結合雙享鍋與壓力鍋兩種優點，無論是煮飯、煮菜、燉湯都在幾分鐘內完成，再難燉煮的食材都可在最短時間萃取，美味極致，讓烹飪既省電又省事，是節能減碳最好的工具。

休閒鍋 HOT PAN

與一般鍋子不同，有內、外鍋之分。內鍋使用不鏽鋼，鍋底能均勻受熱，儲熱性極佳，即使小火就能煮出食物的美味。寬口低身的特殊設計，可當炒鍋、平底鍋及湯鍋使用。擁有二至四小時超強雙重保溫效果，防止食物熱源流失。

UCOM 新都會多層複合金炒鍋單柄

多層複合金不銹鋼的材質，具有快速熱傳導、受熱均勻功能，並有相當高的儲熱功能，不需使用大火烹煮，食物不會沾鍋，還能完整保留食物本身的營養，可搭配蒸籠、蒸盤使用，電磁爐、瓦斯爐皆適用，可滿足多重需求！

雙享鍋

一鍋三用，可當湯鍋、雙享鍋、燜燒鍋。擁有專利雙層鍋蓋與雙層鍋身設計，中空斷熱有效保有鍋內熱源，傳熱速度非常快，六分鐘就能煮好一鍋飯，料理過程可以全程不洗鍋、不放油直接烹調。

攪拌湯杓

一體成型設計，中空斷熱把手，使用不燙手；圓弧設計，連鍋底死角都不放過，鍋底也可輕易撈出，適用於炒菜及拌湯，好清洗不藏污納垢。

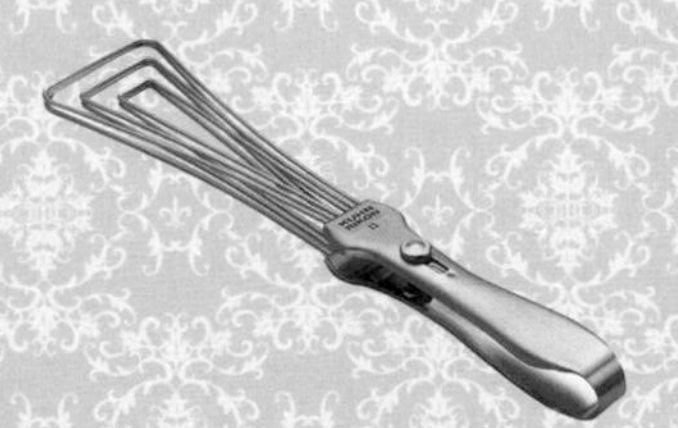

快易夾

可當鍋鏟，攪拌打蛋器，亦可當夾子使用，好握、好洗不傷鍋。

蔬果切片器

雙刃刀片的設計，只需左右擺動食材，即可輕鬆切片，可用於刨小黃瓜、洋芋片、高麗菜絲及洋蔥絲。

易拉轉

免插電的食物調理機，輕輕拉一下，就可以把蔥、薑、蒜、辣椒等辛香料攪細，是廚房不可或缺的小幫手。

玉米轉轉樂

將整隻玉米置於防滑底座內，套上外殼，輕鬆轉一轉，即可隨時隨地吃到新鮮又飽滿的玉米粒。

單柄壓力鍋 3.5L

非常適合都會小家庭使用，此壓力鍋煮白米飯只需三分半鐘，鮮甜入味的滷牛肉也只需要八分鐘即可完成，讓你輕鬆開飯，下廚好愉快。

神奇潔能板

放在瓦斯爐上使鍋子受熱均勻，烹調食物時節省時間，更可以使鍋底不用直接接觸火焰，省去刷洗時間。此外，放置欲退冰的食材於常溫下的節能板，可迅速解凍。

料理的美味秘訣
基本醬汁與高湯

1 奶香白醬 Creamy White Sauce

材料：
奶油 1 大匙、麵粉 3 大匙、牛奶 ½ 杯、無糖鮮奶油 ¼ 杯、鹽適量

做法：
1. 奶油倒入鍋內開小火待融化慢慢倒入麵粉，邊倒入邊攪拌，炒至香味溢出約 8 ～ 10 分鐘，慢慢倒入牛奶拌勻，再倒進鮮奶油拌至無顆粒狀，放入鹽拌勻呈稠狀。

Ingredients:
Butter 1T, Flour 3T, Milk ½ Cup, Sugar-free Cream ¼ Cup, Salt few

Procedure:
1. Melt butter in a pan and add flour slowing while continue stirring (low heat) till savored. It requires around 8 to 10 minutes. Then slowly add milk to blend well. Add whipped cream to blend till smooth. Add salt to blend till thickened.

2 咖哩醬 Western Mixed Soup Stock

材料：
洋蔥末 1 杯、蒜末 2 大匙、薑末 1½ 大匙、奶油 200g、麵粉 450g、咖哩粉 150g

做法：
1. 鍋入奶油待融化倒入洋蔥以小火炒至洋蔥呈茶褐色，倒入咖哩粉（持續小火）炒至咖哩香味溢出（不宜炒過久會有苦味），慢慢放入麵粉，邊炒邊倒入麵粉炒約 10 分鐘。

Ingredients:
Finely-chopped Onion 1 cup, Finely-chopped Garlic 2T, Finely-chopped Ginger 1 ½ T, Butter 200g, Flour 450g, Curry Powder 150g

Procedure:
1. Melt butter in a pan and add onion (low heat) to stir till onion turning brown. Add curry powder (still low heat) to stir till savored (Do not stir too long otherwise it will turn bitter). Slowly add flour to stir for about 10 minutes.

完美烹調寶典 Perfect Cooking Tips

- 可放入冰塊盒，做成咖哩塊存放於冰箱。
- 咖哩使用 3 種以上品牌混合會更香，由於每家咖哩廠的配方不同，所以綜合各家配方可使咖哩粉更加香濃。
- You could put curry paste into ice cube container to make curry paste cube. Store those cubes in refrigerator.
- You could mix more than three brands of curry powder for a strong savor. Since each brand has individual formula, mixing them together will create strong and rich savor.

3 西式高湯 Western Style Soup Stock
（可用於中港台料理）

材料：

雞骨、豬骨各 600g、洋蔥 1 顆（切塊）、西芹 2 枝（切段）、紅蘿蔔 ½ 條（切塊）、月桂葉 1 片、百里香 5 枝（乾燥 ½ 匙）、水 2000cc

做法：

1. 雞骨和豬骨放入烤箱上下火 230℃ 烤至金黃色或入鍋煎。
2. 雞骨、豬骨、洋蔥、西芹、紅蘿蔔、月桂葉、百里香、水、熬煮 5 小時候過濾（若使用壓力鍋熬煮，待上升二條紅線改小火 1 小時）。

Ingredients:

Chicken Bone 600g, Pork Bone600g, Onion 1 (chopped), Salary 2 (segmented), Carrot ½ (chopped), Bally Leaf 1, Thyme 5 (dried ½ T), Water 2 liter

Procedure:

1. Roast chicken bones and pork bones together in oven with top and bottom heat of 230℃ till golden brown or pan fry.
2. Cook the roasted chicken and pork bones, onion, salary, carrot, bally leaf, thyme and braise for 5 hours and then filter.(If using a pressure cooker, you should turn to low heat to cook for 1 hour when it rises to two red strips.)

4 西式牛高湯 Western Style Beef Soup Stock

材料：

牛骨 1.2Kg、水 2500cc、西芹 2 枝（切段）、洋蔥 2 顆（對切）、丁香 8 粒、黑胡椒粒 1 大匙、月桂葉 1 片、百里香 5 枝（或乾燥 ⅓ 大匙）

做法：

1. 牛骨入烤箱烤至呈金黃色。
2. 丁香插入洋蔥放入鍋內，其他材料亦倒入鍋內熬煮 6 ～ 8 小時後過濾（使用壓力鍋熬煮，待上升二條紅線改小火 1.5 小時）。

Ingredients:

Beef Bone 1200g, Water 2500cc, Salary 2 (segmented), Onion 2 (halved),Clove 8, Grained Black Pepper 1T, Bally Leaf 1, Thyme 5 (dried ⅓ T)

Procedure:

1. Roast beef bones in oven till golden brown.
2. Stick cloves into onion and put into pot with other ingredients to braise for 6 to 8 hours. Then filter. (If using a pressure cooker, you should turn to low heat to cook for 1.5hours when it rises to two red strips.)

5 泰式綜合高湯 Thai Style Mixed Soup Stock

材料：

雞骨、豬骨各 600g、蝦殼 300g、香菜頭 5 個、南薑 5 片、香茅 3 枝（切段）、檸檬葉 8 片（撕開）、紅蔥頭 6 粒、水 2500cc

做法：

1. 所有材料放入燉鍋熬煮約 4 ～ 5 小時。使用壓力鍋熬煮待上升二條紅線改小火 1 小時後過濾。

Ingredients:

Chicken Bone 600g, Pork Bone 600g, Shrimp Shells 300g, Coriander Roots 5, Galangal 5 pieces, Lemon Grass 3 (segmented), Lemon Leaf 8 (torn), Shallot 6, Water 2.5 liter

Procedure:

1. Put all ingredients into a pot to braise for 4 to 5 hours. Cook in a pressure cooker and turn to low heat when it rises to two red strips to cook for another 1 hour. Then filter.

6 韓式高湯 Korean Style Soup Stock

材料：
雞骨、豬骨各 600g、昆布 1 大片、洋蔥 1 顆（切塊）、小魚乾 ⅓ 杯、黃豆芽 300g、水 2500cc

做法：
1. 雞骨、豬骨煎至呈金黃色或放入烤箱上下火 220℃烤至呈金黃色。
2. 做法 (1) 放入鍋內並倒入其他材料熬煮 4 ～ 5 小時。
3. 使用壓力鍋熬煮，待上升二條紅線改小火 1 小時後過濾。

Ingredients:
Chicken Bone 600g, Pork Bone600g, Kelp 1 large piece, Onion 1(chopped), Small Dried Fish ⅓ cup, Soy Bean Sprout 300g, Water 2500cc

Procedure:
1. Pan fry chicken and pork bones till golden brown or put into oven to roast with top and bottom heat of 220℃ till golden brown.
2. Put Procedure (1) into a pot with other ingredients to braise for 4 to 5 hours.
3. Cook in a pressure cooker and turn to low heat when it rises to two red strips to cook for another 1 hour. Then filter.

7 日式高湯 Japanese Style Soup Stock

材料：
雞骨、豬骨各 600g、昆布一片、蘋果 1 顆（切塊）、洋蔥 1 顆（切塊）、紅蘿蔔 ½ 條（切塊）、水 2500cc

做法：
1. 雞骨、豬骨汆燙一下或入鍋煎至金黃色。
2. 做法 (1) 放鍋內倒入其他材料（昆布不可洗，以紙巾擦拭）熬煮約 4 ～ 5 小時（使用壓力鍋約煮 1 小時）後過濾。

完美烹調寶典 Perfect Cooking Tips
・可運用煮湯、麵、粥。
・This soup stock is ideal for soup, noodles and congee.

Ingredients:
Chicken Bone 600g, Pork Bone 600g, Kelp 1, Apple 1 (chopped), Onion 1 (chopped), Carrot ½ (chopped), Water 2500cc

Procedure:
1. Blanch chicken and pork bones slightly or pan fry till golden brown.
2. Put Procedure (1) with other ingredients into a pot (Do not rinse kelp. Wipe with paper towel instead) to braise for 4 to 5 hours. (Cook in a pressure cooker would require about 1 hour.) Then filter.

8 越南高湯 Vietnam Style Soup Stock

材料：
雞骨、豬骨各 600g、洋蔥 1 顆（切塊）、香茅 2 枝、小魚乾 1 把、香菜頭 4 個、水 2500cc

做法：
1. 所有材料放入鍋中熬煮約 5 小時（使用壓力鍋約煮 1 小時）後過濾。

Ingredients:
Chicken Bone 600g, Pork Bone 600g, Onion 1 (chopped), Lemon grass 2, Dried Small Fish 1 pinch, Coriander Roots 4, Water2500cc

Procedure:
1. Put all ingredients into a pot to braise about 5 hours. (Cook in a pressure cooker would require about 1 hour) and then filter.

⑨韓式牛高湯 Beef Soup Stock

材料：
牛骨 1.2Kg、昆布 1 大片、洋蔥 1 顆（切塊）、小魚乾 ⅓ 杯、黃豆芽 300g、水 2000cc

做法：
1. 牛骨煎至呈金黃色或放入烤箱以上下火 220℃烤至呈金黃色。
2. 做法 (1) 放入鍋內並倒入其他材料熬煮 6 小時。
3. 使用壓力鍋熬煮，上升二條紅線改小火 1.5 小時後過濾。

Ingredients:
Beef Bone 1200g, Kelp 1 large piece, Onion 1(chopped), Dried Small Fish ⅓ cup, Soy Bean Sprout 300g, Water 2500cc

Procedure:
1. Pan fry beef bones till golden brown or put into oven to roast with top and bottom heat of 220℃ till golden brown.
2. Put Procedure (1) into a pot with other ingredients to braise for 6 hours.
3. Cook in a pressure cooker and turn to low heat when it rises to two red strips to cook for another 1.5 hours. Then filter.

⑩西式綜合高湯 Western Mixed Soup Stock

材料：
雞骨、豬骨各 600g、蝦殼 300g、螃蟹殼 1 隻（切塊）、西芹 2 枝（切段）、紅蘿蔔 ½ 條（切塊）、洋蔥 1 個（切塊）、月桂葉 1 片、百里香 5 枝（或乾燥 1 小匙）、白酒 80cc、胡椒粒 1 大匙、水 2500cc

做法：
1. 蝦跟螃蟹入鍋炒淋上白酒使之蒸發。
2. 雞骨、豬骨移入烤箱上下火 220℃烤至呈金黃色或入鍋煎至呈金黃色。
3. 做法 (1) 及 (2) 放入鍋內並放入其他材料注入水熬煮約 4 ～ 5 小時（壓力鍋上昇二條紅線改小火 1 小時）後過濾。

Ingredients:
Chicken Bone 600g, Pork Bone 600g, Shrimp Shells 300g, Crab Shell 1 (chopped), Salary 1 (segmented), Carrot ½ (chopped), Onion 1 (chopped), Bally Leaf 1, Thyme 5 or dried one 1t, White Wine 80cc, Grain Pepper 1T, Water 2500cc

Procedure:
1. Stir shrimp and crab shells in a stir pan and baste white wine till evaporated.
2. Pan fry beef and chicken bones till golden brown or put into oven to roast with top and bottom heat of 220℃ till golden brown.
3. Put Procedure (1) and (2) and other ingredients into a pot to braise for 4 to 5 hours. (Cook in a pressure cooker and turn to low heat when it rises to two red strips to cook for another 1 hour.) Then filter.

完美烹調寶典 Perfect Cooking Tips

・螃蟹可買較便宜的蚶仔。如無螃蟹亦可。
・此高湯適用於煮湯、麵、粥等。
・熬一大鍋待涼過濾分裝放冷凍。
・You can substitute crab with other cheaper types such as the swimming crabs. Crab shell is optional.
・This soup stock is ideal for soup, noodles and congee, etc.
・Braise a large pot. Filter upon cooling down. Bag the filtered soup stock into small portions to freeze.

瀰漫異國風味的

歐式料理

完美烹調手則

1. 燉飯原文為 Risotto 其意是邊燉煮邊加高湯。
2. 燉飯（義大利米）需使用熱高湯燉煮。
3. 燉飯使用的義大利米，不需洗過，如果米經由洗過，在烹調時高湯就無法被米吸入。
4. 義大利米在燉煮時，要使用木杓不可使用不鏽鋼杓，米粒易被破壞其完整性。
5. 煮義大利麵時水要煮至滾沸麵才會 Q，煮好時以冷水沖涼瀝乾水分拌入 2 大匙橄欖油，分成數份裝入冷凍拉鏈袋置於冷凍，需要時沖熱水即可馬上使用。
6. 料理時，使用乾燥香料的量需比使用新鮮香料的量減少 1 ～ 2 倍。
7. 本書的高湯均可充分利用於火鍋、麵、粥、湯等，也可分裝置冷凍保存。

AMT
AMT

波隆那肉醬義大利麵

Bologna Meat Sauce Spaghetti

材料：

豬絞肉 500g、牛絞肉 500g、培根（切絲）½ 杯、洋蔥 2 顆（切末）、蒜末 2 大匙、番茄糊 3 大匙、番茄粒罐頭 1 罐（切碎）、紅蘿蔔末 ¼ 杯、西芹末 ¼ 杯、西式高湯 3 杯（參見 p.9）、月桂葉 1 片、紅酒 150cc、帕瑪森乳酪適量、西洋香菜少許、義大利麵（煮熟）適量

調味料：

鹽適量、黑胡椒粉適量

做法：

1. 鍋入 1½ 大匙油待溫熱放入培根絲，炒至呈焦黃色即入蒜末爆香至微黃色，倒入洋蔥炒至透亮，再放入紅蘿蔔及西芹炒約 6 分鐘，倒入豬絞肉、牛絞肉炒至肉鬆散，淋上紅酒煮至酒蒸發。
2. 將搗碎的番茄粒和番茄糊倒入做法 (1) 拌勻，注入高湯放進月桂葉，燉煮約 1 小時，最後以適量的黑胡椒及鹽調味。
3. 煮好的義大利麵盛入盤內淋上做法 (2)，撒上帕瑪森乳酪及西洋香菜末。

Ingredient:

Minced Pork 500g, Minced Beef 500g, Bacon (shredded) ½ Cup, Finely-chopped Onion 2, Finely-chopped Garlic 2T, Tomato Paste 3T, Canned Tomato Dice 1 (cut up), Finely-chopped Carrot ¼ Cup, Finely-chopped Salary ¼ Cup, Western Style Soup Stock 3 Cup (please see p.9), Bally Leaf 1 piece, Red Wine 150cc, Parmesan few, Finely-chopped Parsley few, Spaghetti (cooked) some

Seasoning:

Salt and Black Pepper Powder few

Procedure:

1. Add 1½T of oil in a pot to heat and add shredded bacon to stir till golden brown. Add finely chopped garlic to sauté till slightly yellowish. Add onion to stir till transparent. Add carrot and salary to stir about 6 minutes. Add minced pork and beef to stir till loose. Drip red wine to cook till evaporated.
2. Add tomato dices (smashed) and tomato paste in Procedure (1) to mix well. Add soup stock and bally leaf to braise for 1 hour. Lastly add few pinches of black pepper and salt as seasoning.
3. Place the cooked spaghetti in a plate and pour Procedure (2). Sprinkle Parmesan and fine chopped parsley as final garnish.

完美烹調寶典 Perfect Cooking Tips

・蔬菜一定要炒足時間才會釋放出甘甜味。

・Allow enough time to stir vegetables so the nature sweetness will be released.

在所有的義大利麵中這道肉醬義大利麵是最受學生們的家人喜愛，做好的肉醬可運用在焗肉醬千層麵或肉醬焗茄子。

Among many pasta dishes, this meat sauce spaghetti is probably one of the most popular dishes amongst students' family members. The meat sauce in this recipe could be used for lasagne or meat sauce and eggplant au gratin.

米蘭燉牛肉義大利麵

Milan Beef Stew Spaghetti

萬人迷的燉肉義大利麵，口感香濃，回甘香醇是一道必學的義大利麵料理。

The beef stew spaghetti of strong aroma and rich flavor is a real charmer. The mellow and savory aftertaste makes this dish a must-learn pasta dish.

材料：

牛肋條（牛腩）1kg、培根 5 片（切 0.5cm 寬）、洋蔥末 1 杯、紅蘿蔔末 ¼ 杯、西芹末 ¼ 杯、蒜末 3 大匙、紅酒 120cc、番茄粒罐頭 1 罐（切碎）、番茄配司 1½ 大匙、月桂葉 1 片、西洋香菜少許、帕瑪森乳酪適量、義大利麵適量、西式牛高湯或西式高湯 1 杯（參見 p.9）、麵粉適量

調味料：

黑胡椒粉適量、鹽適量

醃料：

黑胡椒粉適量、鹽適量，黑胡椒粉與鹽以 2：1 的比例拌勻。

做法：

1. 牛肋條切塊狀撒上醃料放置 10 分鐘後沾上麵粉。
2. 鍋入 1½ 大匙油待熱放入做法 (1) 煎至呈焦黃色取出備用；將培根倒入鍋內煎炒至焦黃色，放入蒜末爆香即入洋蔥末以小火炒至呈透明狀，再倒入紅蘿蔔末，西芹末約炒 8 分鐘後改大火放入煎好的牛肋條，淋上紅酒使之蒸發，並倒入番茄粒碎塊及番茄配司拌勻，注入高湯放入月桂葉蓋上鍋蓋，待冒煙改小火約 60 分鐘（如材質較差的鍋子需要增加 20 分鐘熬煮，必須再增加 1½ 杯的高湯）。使用壓力鍋待上升二條紅線改小火 22 分鐘，最後以黑胡椒粉及鹽調味。
3. 水煮沸放入鹽 2 大匙、橄欖油 2 大匙倒入義大利麵約煮 8 分鐘，取出瀝乾水分放入器皿內，淋上做法 (2) 撒上帕瑪森乳酪及西洋香菜。

Ingredient:

Beef Rib Fingers (Beef Tenderloin) 1kg, Bacon 5 pieces(Cut into 0.5cm), Finely-chopped Onion 1 Cup, Finely-chopped Carrot ¼ Cup, Finely-chopped Salary ¼ Cup, Finely-chopped Garlic 3T, Red Wine 120 cc, Canned Tomato Dices 1 (cut up), Tomato Paste 1 ½ T, Bally Leaf 1 piece, Parsley few, Parmesan some, Spaghetti some, Western Style Soup Stock or Western Style Beef Soup Stock 1 Cup (please see p.9), Flour some

Seasoning:

Black Pepper Powder some, Salt some

Marinate:

Black Pepper Powder some, Salt some.The proportion between black pepper powder and salt is 2:1.

Procedure:

1. Cut beef rib finger into chucks and sprinkle with marinate ingredients. Leave for 10 minutes and then coat with flour.
2. Add 1 ½ T of oil in a pot to pan-fry Procedure (1) till golden brown. Take out and set aside; Add bacon into this pot to pan-fry till golden brown. Add finely chopped garlic to sauté and then add finely chopped onion to stir till transparent (low heat). Add finely chopped carrot and salary to stir about 8 minutes and then turn to high heat. Add the pan-fry beef rib finger and drip red wine to cook till evaporated. Add tomato dices and tomato paste to blend evenly. Add soup stock and bally leaf, cover with lid, and turn to low heat upon smoking to cook about 60 minutes (If using pots of less ideal materials, it requires extra 20 minutes to cook. Then add 1 ½ cups of soup stock.). If using a pressure cooker, turn to low heat to cook for 22 minutes when it rises to two red strips. Lastly add black pepper powder and salt as seasoning.
3. Boil water and add 2 tablespoons of salt and 2 tablespoons of olive oil. Add spaghetti to cook for 8 minutes. Take out and drain off water. Place the cooked spaghetti in a container, pour Procedure (2) and sprinkle Parmesan and parsley as final garnish.

完美烹調寶典 Perfect Cooking Tips

- 可改選用豬梅花肉或牛腱肉。
- 蔬菜一定要炒至時間夠才會釋出甘醇味。
- 可搭配烤得香酥的法國麵包。

- You could substitute with pork shoulder or beef shank.
- Allow enough time to stir vegetable so that the nature sweetness would be released.T
- This recipe could also serve with baked baguette.

南瓜鮮蝦義大利麵

Pumpkin and Shrimps Fettuccine

材料：

義大利寬麵 300g、鮮蝦 300g、南瓜 400g、鮮奶油 120cc、洋蔥末 ½ 杯、蒜末 ½ 大匙、奶油 2 ½ 大匙、帕瑪森乳酪適量、蘆筍 150g、西式綜合高湯 ½ 杯（參見 p.11）、白酒 60cc

調味料：

鹽適量

做法：

1. 蝦去頭去殼留尾巴，取出腸泥；南瓜去皮去籽切小丁。
2. 鍋入 1 ½ 大匙奶油待融化，放入洋蔥末（留 1 大匙備用）炒透呈透明狀，倒入南瓜炒勻，注入高湯煮至南瓜熟爛，過篩成泥狀（或入調理機打成泥狀）倒入鍋內。
3. 鮮奶油倒入做法 (2) 內拌勻煮一下以適量的鹽調味。
4. 義大利寬麵倒入已加鹽及橄欖油的沸水中煮熟取出瀝掉水分。
5. 蘆筍汆燙熟取出斜切段。
6. 鍋入 1 大匙奶油待融化放入蒜末爆香，再入洋蔥末炒香，放入蝦撒上白酒拌勻，倒入做法 (3) 及 (4)，以適量鹽調味拌勻，再放入做法 (5) 盛入器皿內，撒上帕瑪森乳酪。

Ingredient:

Fettuccine 300g, Shrimp 300g, Pumpkin 400g, Cream 120cc, Finely-chopped Onion ½ Cup, Finely-chopped Garlic ½ T, Butter 2 ½ T, Parmesan few, Asparagus 150g, Western Mixed Soup Stock ½ Cup (please see p.11), White Wine 60cc

Seasoning:

Salt few pinches

Procedure:

1. Shell shrimps with tails left. Remove heads and intestine mud. Peel and seed the pumpkin and cut into dices.
2. Melt 1 ½ T of butter in a pot and add finely chopped onion (leave 1T) to stir till transparent. Add pumpkin to stir and blend well. Add soup stock to cook till pumpkin tendered. Sieve into paste or put into juicer to mix into paste. Add into a pot.
3. Add cream into Procedure (2) to slightly blend and cook. Add few pinches of salt as seasoning.
4. Add fettuccine into boiling water (with salt and olive oil) to cook till done. Take the cooked fettuccine out and drain off water.
5. Blanch asparagus till cooked. Take the blanched asparagus out and cut into segments.
6. Melt 1T of butter in a pot and add finely chopped garlic to sauté. Add finely chopped onion to sauté, Add shrimps, drip white wine and blend well. Add Procedure (3) and Procedure (4). Sprinkle Parmesan.

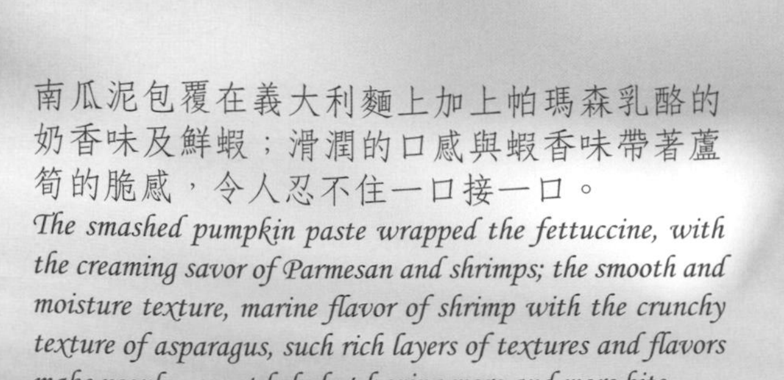

南瓜泥包覆在義大利麵上加上帕瑪森乳酪的奶香味及鮮蝦；滑潤的口感與蝦香味帶著蘆筍的脆感，令人忍不住一口接一口。

The smashed pumpkin paste wrapped the fettuccine, with the creaming savor of Parmesan and shrimps; the smooth and moisture texture, marine flavor of shrimp with the crunchy texture of asparagus, such rich layers of textures and flavors make people cannot help but having more and more bite.

牛肝蕈野菇雞丁義大利麵

Porcini Mushroom, Wild Mushroom and Chicken Dice with Fettuccine

材料：

乾燥牛肝菌 30g、鮮香菇 8 朵（切條狀）、蘑菇 100g（切片）、義大利寬麵（煮熟）適量、去骨雞腿 2 隻、洋蔥末 3 大匙、蒜末 1 大匙、帕瑪森乳酪適量、白酒 80cc、西式高湯 ⅕ 杯（參見 p.9）、西洋香菜少許

調味料：

黑胡椒粉適量、鹽適量

醃料：

黑胡椒粉與鹽的比例為 2：1、香蒜粉 ½ 小匙、匈牙利紅椒粉 ½ 小匙、檸檬汁 ½ 大匙

做法：

1. 乾燥牛肝菌浸泡溫水 10 分鐘後取出切丁，湯汁當高湯使用。
2. 雞肉切塊先淋上檸檬汁拌勻，再撒上其他醃料放置 30 分鐘後，放入鍋內煎至熟呈金黃色取出備用。
3. 油 1 大匙倒入做法 (2) 的鍋內，待熱倒入蒜末爆炒至呈微黃色，再入洋蔥末炒香，即入牛肝菌爆香再倒入鮮香菇、蘑菇片拌炒，淋上白酒，放入煎熟的雞塊淋上高湯及牛肝菌（浸泡）汁，放入煮熟的義大利麵，快炒幾下；以適量的黑胡椒粉、鹽調味，並撒上 1 大匙的帕瑪森乳酪拌勻，盛入盤內撒上帕瑪森乳酪及西洋香菜末。

Ingredient:

Dried Porcini Mushroom 30g, Fresh Shitake Mushroom 8 (cut into strips), Mushroom 100g (sliced), Cooked Fettuccine some, Boned Drumstick 2, Finely-chopped Onion 3T, Finely-chopped Garlic 1T, Parmesan few, White Wine 80cc, Western Style Soup Stock ⅕ Cup (Please see p.9), Finely-chopped Parsley few

Seasoning:

Black Pepper Powder few, Salt

Marinate:

Black Pepper Powder with Salt (proportion is 2:1), Garlic Powder ½ t, Hungarian Red Pepper Powder ½t, Lemon Juice ½ T

Procedure:

1. Soak dried porcini mushroom in warm water for 10 minutes. Take the soaked porcini mushroom out and cut in dices. The soak water is used as soup stock.
2. Cut chicken into chucks and blend with lemon juice. Add other marinate ingredients and leave for 30 minutes. Pan-fry till golden brown and take the pan-fried chicken out. Set aside.
3. Add 1T of oil in the pot of Procedure (2) to heat and add finely chopped garlic to sauté till slightly yellowish. Add finely-chopped onion to sauté. Then add porcini mushroom to sauté and add fresh shitake mushroom and sliced mushroom to stir slightly. Drip white wine. Add pan-fried chicken chucks, soup stock and the porcini mushroom soaking juice. Add cooked Fettuccine to quickly stir. Add few pinches of black pepper powder and salt as seasoning. Sprinkle 1T of Parmesan to blend well. Place to a plate and sprinkle Parmesan and finely chopped parsley as final garnish.

完美烹調寶典
Perfect Cooking Tips

・吃素者，以杏鮑菇切片代替雞肉，高湯以鮮奶油代替，不放洋蔥及蒜末，放 40g 牛肝菌。

・For vegetarian, substitute chicken with king oyster mushroom, soup stock with cream, no finely chopped onion and garlic. Add 40g of porcini mushroom instead.

牛肝蕈的香氣搭配著野菇的口感更增添其義大利麵的迷人風采。

The aroma of porcini mushroom combining with the rich texture of wild mushroom makes this pasta dish a very charming and pleasant dish.

艾米利亞海鮮焗千層麵

Emilia Seafood Lasagne

材料：

義大利千層麵適量、蝦仁 200g、白魚肉 200g（去骨）、蛤蠣 300g、透抽 200g、蟹肉 100g、洋蔥末 3 大匙、蒜末 1 大匙、白酒 100cc、西芹末 1½ 大匙、莫查瑞拉乳酪適量、帕瑪森乳酪適量、西洋香菜少許適量、麵包粉 1 大匙、奶香白醬 2 杯（參見 p.8）、豆蔻粉 ¼ 小匙

做法：

1. 千層麵皮放入已加入鹽及橄欖油的沸水內煮熟後取出沖冷水，冷卻後瀝乾水分。
2. 奶香白醬放入少許荳蔻粉拌勻。
3. 鍋內放入 ½ 大匙油放入蒜末、洋蔥末爆香，放入西芹末，再放入蛤蠣略炒幾下，倒入其他海鮮，淋上白酒以少許黑胡椒粉、鹽調味，盛出後取出蛤蠣殼。
4. 將做法 (2) 及做法 (3) 拌勻。
5. 器皿內塗上少許奶油鋪上做法 (4) 放上一張千層麵，撒上適量的莫查瑞拉乳酪，再鋪上做法 (4) 依續重疊 3 次，最後置上一張千層麵，鋪上莫查瑞拉乳酪，撒上帕瑪森乳酪，再撒上麵包粉放入烤箱以上火 230℃ 烤至呈焦黃色即可。

Ingredient:

Italian Lasagne some, Shelled Shrimp 200g, White Fish Filet 200g(boned), Clam 300g, Squid 200g, Crab Meat 100g, Finely-chopped Onion 3T, Finely-chopped Garlic 1T, White Wine 100cc, Finely-chopped Salary 1 ½ T, Mozzarella Cheese few, Parmesan some, Finely-chopped Parsley few, Bread Powder 1T, Creamy White Sauce 2 Cups (please see p.8), Nutmeg Powder ¼ t

Procedure:

1. Cook lasagne in boiling water (with salt and olive oil) till done and take the cooked lasagna out. Cool down with running water and then drain off water.
2. Add nutmeg powder in creamy white sauce and mix well.
3. Add ½ T of oil in a pot and add finely chopped garlic and onion to sauté. Add finely chopped salary with clam to stir slightly. Add other seafood, drip white wine, add few pinches of black pepper powder and salt as seasoning. Take out and remove the clam shells.
4. Blend Procedure (2) and Procedure (3) evenly.
5. Spread butter insider the container and put the Procedure (4) on bottom with a piece of lasagne covered. Sprinkle some Mozzarella cheese. Then add Procedure (4) again with another piece of lasagne covered and sprinkle Mozzarella cheese. Repeat three times with this said order. Lastly put a piece of lasagne and Mozzarella cheese, sprinkle Parmesan and some bread powder. Put into oven to bake with top heat of 230℃ till golden brown..

CC的學生和朋友們常常問CC為什麼外面餐廳焗類料理特別偏油膩？希望CC能多教些焗的料理；其實焗的重點在於白醬中的麵粉炒的好不好，今天就跟著CC做看看！

Many CC's students and friends often ask why the gratin dishes at restaurants are always very greasy. They thus hope CC could teach more gratin dishes; in fact, the key point to good gratin dishes is how the flour of white sauce is stirred. Today you can learn to do it with CC!

義大利辣味茄汁海鮮麵

Italian Spicy Tomato Sauce Seafood Spaghetti

材料：

蝦 8 尾、鮮干貝 4 個、淡菜 4 個、紅醬 ½ 杯、TABASCO 醬適量、洋蔥末 3 大匙、蒜末 1 大匙、紅辣椒末 1 大匙、白酒 80cc，義大利麵適量（煮熟）、羅勒適量

紅醬：

番茄粒罐頭 1 罐（搗碎）、洋蔥末 ¼ 杯、蒜末 1 大匙、西式綜合高湯 ⅔ 杯（參見 p.11）、月桂葉 1 片、黑胡椒粉適量、鹽適量、俄力崗香料 ½ 小匙

紅醬做法：

鍋入 1 大匙油待熱倒入蒜末爆香，再放入洋蔥末炒至呈透明狀，倒入番茄粒末、高湯、月桂葉及俄力崗香料熬煮約 15 分鐘，最後以適量的鹽及黑胡椒粉調味。

做法：

1. 蝦由背部劃開取出腸泥。
2. 鍋入 1 大匙油待熱放入蒜末爆炒至呈微黃色，再倒入洋蔥末和紅辣椒末炒香，放進蝦、淡菜及干貝略炒幾下，淋上白酒煮，倒入紅醬放入煮好的義大利麵、適量的 TABASCO 和適量的鹽及黑胡椒粉，快炒幾下撒上羅勒盛入器皿內。

Ingredient:

Shrimp 8, Raw Scallop 4, Mussel 4, Red Sauce ½ Cup, TABASCO sauce some, Finely-chopped Red Onion 3T, Finely-chopped Garlic 1T, Finely-chopped Red Chili Pepper 1T, White Wine 80cc, Cooked Spaghetti some, Basil Leaf some

Red Sauce:

Canned Tomato Dice 1(smashed), Finely-chopped Onion ¼ Cup, Finely-chopped Garlic 1T, Western Mixed Soup Stock ⅔ Cup (please see p.11), Bally Leaf 1 piece, Black Pepper Powder few, Salt few, Oregano Spice ½ t

Procedure: of Red Sauce:

Add 1T of oil in a pot and add finely-chopped garlic to sauté. Add finely chopped onion to stir till transparent. Add tomato dices, soup stock, bally leaf and oregano spice to braise for 15 minutes. Finally add some pinches of salt and black pepper powder as seasoning.

Procedure:

1. Cut the back of shrimps to remove the intestine mud.
2. Add 1T of oil in a pot to heat and add finely chopped garlic to sauté till slightly yellowish. Add finely chopped onion and red chili pepper to sauté. Add shrimps, mussels and scallops to stir gently. Drip white wine to cook and then add red sauce and cooked spaghetti with TABASCO sauce, salt and black pepper powder to stir quickly. Sprinkle basil leaf and place to a container to serve.

完美烹調寶典
Perfect Cooking Tips

‧紅醬可多做一些分裝好放入冷凍可存放一個月。

‧You could make more red sauce and bag it separately. It could last for one month if frozen in refrigerator.

CC和好朋友們到義大利餐廳的首選麵，但是每次都吃不過癮，太貴了！自己做最實在！只要自己會熬煮紅醬一切就完美了！

The priority choice of pasta for CC and close friends at any Italian restaurants is this dish all the time. However, it is always not enough and so pricy as well. It is better to do it at home, so long as you learn how to make red sauce. It is such perfect solution!

朵倫諾牛肉焗通心麵（筆管麵）

Torino Beef and Penne au Gratin

材料：

牛肋條（牛腩）1kg、洋蔥末 1 杯、蒜末 2 大匙、培根 5 片（切 0.5cm 寬）、紅酒 150cc、蘑菇 150g（切片）、月桂葉 1 片、番茄粒罐頭（切碎）、番茄醬 2 大匙、筆管麵 ½ 包（煮熟）、莫查瑞拉乳酪適量、帕瑪森乳酪適量、西洋香菜少許、麵粉適量、奶香白醬 1 杯（參見 p.8）、牛高湯 3 杯（參見 p.11）

醃料：

黑胡椒粉適量、鹽適量、紅椒粉 1 小匙

做法：

1. 牛肋條切長約 5 公分，撒上醃料放置 15 分鐘後沾上麵粉入鍋煎至呈金黃色取出。
2. 1 大匙油倒入做法 (1) 的鍋內放入蒜末爆香，再入洋蔥末炒至透明狀，倒入蘑菇片略炒幾下再倒入做法 (1) 淋上紅酒使之蒸發，放入番茄粒及番茄配司拌匀，淋上高湯及放入月桂葉燉煮約 1 小時，以黑胡椒粉及適量的鹽調味拌勻。
3. 筆管麵放於調理盆內倒入適量的鹽拌勻，再放入適量的做法 (2) 及奶香白醬和 ½ 杯乳酪絲拌勻。
4. 做法 (3) 倒入器皿內舖上莫查瑞拉乳酪撒上帕瑪森乳酪，移入烤箱以上火 230℃ 烤至金焦黃色取出撒上西洋香菜。

Ingredient:

Beef Rib Finger (Beef Tenderloin) 1kg, Finely-chopped Onion 1 Cup, Finely-chopped Garlic 2T, Bacon 5 pieces (cut 0.5cm), Red Wine 150cc, Mushroom(sliced)150g, Bally Leaf 1 piece, Canned Tomato Dices 1 (cut up), Ketchup 2T, Cooked Penne ½ bag, Mozzarella Cheese some, Parmesan some, Parsley few, Flour some, Creamy White Sauce 1 Cup (please see p.8), Beef Soup Stock 3 Cups (please see p.11)

Marinate:

Black Pepper Powder some, Salt some, Red Pepper Powder 1t

Procedure:

1. Cut beef rib finger in 5cm long. Sprinkle the marinate ingredients and leave for 15 minutes. Dip flour and pan-fry till golden brown. Take the pan-fried beef out.
2. Add 1T of oil in the pot of Procedure (1) and add finely chopped garlic to sauté. Then add finely chopped onion to stir till transparent. Add sliced mushroom to stir slightly and then add Procedure (1). Drip red wine to cook till evaporated. Add tomato dices and tomato paste to blend well. Pour soup stock and add bally leaf to braise for 1 hour. Add black pepper powder and salt as seasoning.
3. Add penne in a basin and add few pinches of salt to mix well. Add some Procedure (2), creamy white sauce and ½ cup of shredded cheese to blend evenly.
4. Place Procedure (3) in a container, top with Mozzarella cheese and sprinkle Parmesan. Put into oven to bake with top heat of 230℃ till golden brown. Take out and sprinkle parsley as final garnish.

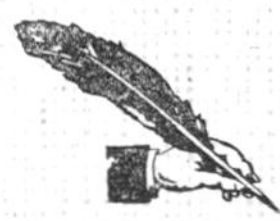

完美烹調寶典 Perfect Cooking Tips

- 做法 (2) 可當成一道義大利燉肉料理。
- 可改用牛腱肉或豬梅花肉。
- 紅酒一定要蒸發完全才不會酸澀。

- Procedure: (2) could be one solid Italian stew dish.
- You could substitute with beef shank or pork shoulder.
- Red wine must be totally evaporated so that there won't be sour taste.

最喜歡做這道料理了，因為用途多變，只要將做法 *(2)* 完成即成為一道燉肉料理，可搭配法國麵包或是以飯為基底作為牛肉焗飯；其受歡迎的程度是只要每位學習過的學員都會說這道料理已成為他們的拿手招牌菜。

I really love to make this dish because it is a wonderful base for many other dishes. You would get a delicious beef stew by finishing Procedure (2) and you could serve with Banquette or top over rice as beef stew rice bowl; you can tell how popular this dish is by the fact that all students who learned to cook it said that this is their excellent dishes.

西班牙漁夫海鮮麵

Spanish Fisherman Seafood Spaghetti

這是西班牙漁夫最拿手的料理哦！

This is one of the most famous specialty dishes by Spanish fisherman!

材料：

義大利麵 300g、蝦 200g、透抽 1 隻、海瓜子或蛤蠣 300g、貽貝 8 個、石狗公魚 1 尾、洋蔥末 1 杯、紅辣椒末 1 大匙、蒜末 2 大匙、西式綜合高湯 ½ 杯（參見 p.11）、番茄粒罐頭 1 罐、白酒 100cc、羅勒葉末 1 大匙、西洋香菜末適量、百里香末 ½ 大匙、俄力崗香料 ½ 小匙

調味料：

黑胡椒粉及鹽適量

做法：

1. 石狗公撒上適量的鹽及黑胡椒粉放置 15 分鐘，透抽切圈狀。
2. 鍋內倒入 1 大匙油放入蒜末爆香，即入洋蔥末炒至呈透明狀，倒入搗碎的罐頭番茄粒，注入高湯放入俄力崗、百里香和羅勒葉末約煮 10 分鐘後，以適量的鹽及黑胡椒粉調味。
3. 鍋內入 1 大匙油待熱放入做法 (1) 的石狗公煎熟取出備用。
4. 沸水內放入 1½ 大匙鹽及 2 大匙橄欖油，放入義大利麵約煮 8 分鐘，撈出瀝掉水分。
5. 鍋倒入 1 大匙油放入蒜末炒至微黃，即入洋蔥末爆香，並放入紅辣椒末倒入蛤蠣約炒 1 分鐘，放入貽貝、蝦及透抽淋上白酒煮至蒸發後，倒入做法 (2) 再倒入做法 (4) 拌炒勻至海鮮熟，放入適量的調味料拌勻，鋪上煎好的魚再撒上羅勒及西洋香菜即可。

Ingredient:

Spaghetti 300g, Shrimp 200g, Squid 1, Variegate Venus or Clam 300g, Mussel 8, Red Rock Cod 1, Finely-chopped Onion 1 Cup, Finely-chopped Red Chili Pepper 1T, Finely-chopped Garlic 2T, Western Mixed Soup Stock ½ Cup (Please see p.11), Canned Tomato Dice 1, White Wine 100cc, Finely-chopped Basil Leaf 1T, Finely-chopped Parsley few, Finely-chopped Thyme ½ T, Oregano Spice ½ t

Seasoning:

Black Pepper Powder and Sale few

Procedure:

1. Sprinkle few pinches of salt and black pepper powder onto the red rock cod and leave for 15 minutes. Cut the squid into rings.
2. Add 1T of oil in a pot and add finely-chopped garlic to sauté. Add finely chopped onion to stir till transparent. Add the canned tomato dices (smashed), soup stock, oregano spice and basil leafs to cook for 10 minutes. Add few pinches of salt and black pepper powder as seasoning.
3. Add 1T of oil to heat and add Procedure (1) Red Rock Cod to pan-fry. Take the pan-fried Procedure (1) out and set aside.
4. Add 1 ½ T of salt and 2T of olive oil in boiling water. Add spaghetti to cook about 8 minutes. Take the cooked spaghetti out and drain off water.
5. Add 1T of oil in a pot and add finely chopped garlic to stir till slightly yellowish. Add finely chopped onion to sauté. Add finely chopped red chili pepper and clam to stir about 1 minute. Add mussel, shrimp and squid and drip white wine to cook till evaporated. Add Procedure (2) first and then add Procedure (4) to blend well till seafood cooked. Add seasoning to mix evenly. Put the pan-fried fish on top and sprinkle basil leaf and parsley as final garnish.

・煮義大利麵的熟度視各廠牌的標示為主，大致上約煮 8 ～ 12 分鐘。

・The degree of cooking for pasta is subject to different brands and as indicated on the instruction on label. Approximately it requires 8 to 12 minutes.

鮭魚白酒義大利麵

Salmon with White Wine Spaghetti

材料：

鮭魚 600g、白酒 120cc、檸檬汁 1½ 大匙、義大利麵 300g、無糖鮮奶油 ⅔ 杯、檸檬皮末 1 大匙、西式綜合高湯 3 大匙（參見 p.11）

調味料：

鹽適量

醃料：

橄欖油 1½ 大匙、鹽適量、檸檬汁 1 大匙

做法：

1. 鮭魚切 1cm 厚，淋上橄欖油及檸檬汁，撒上適量的鹽放置 20 分鐘。
2. 沸水內放入 2 大匙橄欖油及鹽，放入義大利麵煮熟取出瀝掉水分。
3. 鍋入做法 (1) 煎至 9 分熟，撒上白酒倒入鮮奶油，煮一下取出鮭魚備用，倒入做法 (2) 以適量的鹽調味，並淋上檸檬汁快速拌勻起鍋，盛入器皿內放上鮭魚，撒上檸檬皮末。

Ingredient:

Salmon Filet 600g, White Wine 120cc, Lemon Juice 1 ½ T, Spaghetti 300g, Sugar-free Cream ⅔ Cup, Lemon Zest 1T, Western Mixed Soup Stock Soup Stock 3T (please see p.11)

Seasoning:

Salt few pinches

Marinate:

Olive Oil 1 ½ T, Salt some, Lemon Juice 1T

Procedure:

1. Cut salmon filet into 1cm thick and drip olive oil, lemon juice and few pinches of salt. Leave for 20 minutes.
2. Add 2T of olive oil and salt in boiling water to cook spaghetti till done. Take the cooked spaghetti out and drain off water.
3. Add Procedure (1) in a pot to pan-fry till almost done. Drip white wine and cream to slightly cook. Take the salmon filet out and set aside. Add Procedure (2) with few pinches of salt as seasoning. Drip lemon juice to stir quickly before turning off the heat. Place in a container and add salmon at top. Sprinkle lemon zest as final garnish.

完美烹調寶典
Perfect Cooking Tips

- 檸檬皮不可削至白色部分，品嘗時會有苦澀味。
- 可使用煙燻鮭魚代替。

- Lemon zest should not get the white skin part otherwise it will have bitter taste.
- You could substitute with smoked salmon.

檸檬的清香味讓整個義大利麵更增添其迷人的風味。
The citrus freshness of lemon accents the enchanting flavor of this spaghetti dish.

第凡內鮮蝦燴飯

Tiffany Shrimp Gravy on Rice

材料：

蝦 600g、米 2 杯、蘑菇 150g（切片）、蒜末 1 大匙、白蘭地 2 大匙、白酒 200cc、西式高湯 2 杯（參見 p.11 頁）、洋蔥末 3 大匙、奶油 2½ 大匙、匈牙利紅椒粉 ⅓ 小匙、奶香白醬 1 杯（參見 p.8 頁）

調味料：

鹽適量

做法：

1. 鍋入奶油 1½ 大匙待融化，放入蒜末及洋蔥末 2 大匙爆香，即入蘑菇略炒幾下淋上白酒 120cc 炒勻，再倒入米以中小火炒約 2 分鐘，注入高湯放入適量的鹽拌勻，蓋上鍋蓋待冒煙改小火約煮 25 ～ 30 分鐘（使用休閒鍋待冒煙改小火煮 6 分鐘熄火，移入外鍋燜 6 分鐘）。
2. 蝦去殼留尾巴，背部劃開取腸泥。
3. 奶油 1 大匙放入鍋內待融化，放入 1 大匙洋蔥末爆香，即放入蝦淋上白蘭地煮至酒蒸發，再淋上 80cc 白酒煮一下取出蝦備用。
4. 奶香白醬放入鍋內撒上匈牙利紅椒粉以打蛋器拌勻，放入做法 (3) 拌勻，以少許鹽調味即可。
5. 做法 (1) 盛入器皿內再放上做法 (4) 即可。

Ingredient:

Shrimp 600g, Rice 2 Cups, Mushroom 150g (sliced), Finely-chopped Garlic 1T, Brandy 2T, White Wine 200cc, Western Mixed Soup Stock 2 Cups (please see p.11), Finely-chopped Onion 3T, Butter 2 ½ T, Hungarian Red Pepper Powder ⅓ t, Creamy White Sauce 1 Cup (please see p.8)

Seasoning:

Salt few pinches

Procedure:

1. Melt 1 ½ T of butter in a pot and add 2T of finely chopped garlic and onion to sauté. Add mushrooms to slightly stir and drip 120cc of white wine to blend well. Add rice to stir for 2 minutes with low heat. Add soup stock with some pinches of salt to mix evenly. Cover with lid to cook till smoking. Turn to low heat to cook for another 25 to 30 minutes. (If using HotPan, turn to low heat to cook for 6 minutes upon smoking. Then remove to outer cooker to simmer for 6 minutes.)
2. Shell the shrimps with tails left. Cut from the back of shrimps to remove the intestine mud.
3. Melt 1T of butter in a pot and add 1T of finely chopped onion to sauté. Add shrimps and drip brandy to cook till brandy evaporated. Then drip 80cc of white wine to slightly cook. Take the shrimps out and set aside.
4. Add creamy white sauce in a pot and sprinkle Hungarian red pepper powder to blend evenly (with egg-whisker). Add into Procedure (3) to mix well. Add few pinches of salt as seasoning.
5. Place Procedure (1) in a container and then add Procedure (4).

完美烹調寶典
Perfect Cooking Tips

- 使用匈牙利紅椒粉可使顏色更加紅粉，類似蝦膏的色澤。

- The Hungarian red pepper powder could bring out rosy red hue, which is very similar to shrimp paste.

這道是 CC 的懶人宴客料理，非常受客人的喜愛，在此與大家分享。

This dish is CC's lazy feast dish, which is easy, quick and fond by guests. CC shares the secret recipe with readers here.

加州香料烤雞腿鍋飯

California Spice Roasted Drumstick Casserole Rice

材料：

去骨雞腿 3 隻、米 1½ 杯、西式高湯 1½ 杯（參見 p.9）、奶油 1 大匙、蒜末 1 大匙、迷迭香 2 枝、蒜頭 10 顆

醃料 A：

蒜末 1½ 大匙、巴沙米可醋 1½ 大匙、橄欖油 1½ 大匙、檸檬汁 1 大匙、迷迭香末 1½ 大匙、西洋香菜 ¼ 大匙，拌勻

醃料 B：

黑胡椒粉及鹽適量、香蒜粉 ¼ 大匙、紅椒粉適量

做法：

1. 雞腿瀝乾水分撒上醃料 B，再拌入醃料 A 放置一個晚上是最美味的，如果急用至少需放置 2 個小時。
2. 鍋入奶油待融化放入蒜末爆香，倒入米略炒幾下，注入高湯放入適量的鹽拌勻，約煮 23 ～ 25 分鐘煮至米熟。
3. 鍋預熱放入做法 (1) 皮先朝下約 1 分鐘後翻面，移入烤箱以上下火 220℃ 烤，並鋪上 2 枝迷迭香及蒜頭約烤 15 ～ 20 分鐘（視烤箱廠牌）或在鍋內煎烤至熟。
4. 取一只小鍋放於爐火上倒入做法 (2)，鋪上做法 (3) 將整鍋取出放置盤中。

Ingredient:

Boned Drumstick 3, Rice 1 ½ Cup, Western Style Soup Stock 1 ½ Cup (please see p.9), Butter 1T, Finely-chopped Garlic 1T, Rosemary 2, Garlic 10

Marinate: A

Finely-chopped Garlic 1 ½ T, Balsamic Vinegar 1 ½ T, Olive Oil 1 ½ T, Lemon Juice 1T, Finely-chopped Rosemary 1 ½ T, Finely-chopped Parsley ¼ T. Blend all ingredients well.

Marinate: B

Black Pepper Powder and Salt some. Garlic Powder ¼ T, Red Pepper Powder some

Procedure:

1. Drain the drumstick off moisture and sprinkle seasoning B. Then blend with seasoning A well and leave for overnight (the most delicious way). If in rush, at least leave for 2 hours.
2. Melt butter in a pot and add finely chopped garlic to sauté. Add rice to stir slightly and then add soup stock with few pinches of salt to blend evenly. Cook for 23 to 25 minutes till rice cooked.
3. Heat a pan and add Procedure (1) with skin side facing down to pan-fry 1 minute and then turn the other side down. Put into oven with both top and bottom heat of 220° to roast with 2 branches of rosemary and garlic for 15 to 20 minutes (depending on the brand of oven) or directly pan-fry in the pan till cooked.
4. Take one small pot or casserole on stove and add Procedure (2) with Procedure (3) on top. Take the whole pot off stove and place in a plate to serve.

完美烹調寶典
Perfect Cooking Tips

・建議料理時使用新鮮的迷迭香，口感及香味較佳。

・It is suggested to use fresh rosemary when cooking for better texture and savor.

這道烤雞腿飯是 CC 在舊金山的義大利城學到的，它融合了義大利及當地加州菜的特色，CC 再加以改成國人可以接受的口感，大受稱讚。

This roasted drumstick casserole rice is learnt at the Italian Town of San Francisco. It infuses Italian cuisine with the local California style. CC further adjusts it to the flavor accepted by Taiwanese people and this dish is widely proclaimed.

巴黎式海鮮焗飯

Parisian Seafood & Rice au Gratin

材料：

米 2 杯、西式綜合高湯 2 杯（參見 p.11）、奶油 4 大匙、蟹肉或蟹管肉 150g、蝦 200g、貽貝 8 個、透抽 1 隻、蘑菇 12 朵（切片）、洋蔥 ½ 顆（切末）、帕瑪森乳酪適量、蛋黃 2 顆、起司 160g、白酒 80cc、牛奶 1 杯、麵粉 4 大匙、西洋香菜末少許

做法：

1. 1½ 大匙奶油放入鍋內，待融化放入洋蔥末（留 1 大匙備用）炒香，即入蘑菇片略炒幾下，倒入米以小火炒約 1 分鐘，注入高湯蓋上鍋蓋待冒煙改小火，一般鍋約煮 25～30 分鐘（使用休閒鍋待冒煙改小火煮約 6 分鐘，熄火移外鍋燜 6 分鐘）。
2. 2 大匙奶油放入鍋內待融化倒入麵粉以小火炒香，炒約 8 分鐘倒入牛奶拌勻至無顆粒狀，放入 40g 的起司和適量的鹽拌勻。
3. 透抽切圈狀，蝦去殼留尾巴。
4. 鍋內放入 ½ 大匙奶油，待融化放入 1 大匙洋蔥爆香，放入海鮮略炒幾下，淋上白酒使之蒸發，以適量的黑胡椒粉、鹽調味。
5. 將做法 (1) 倒入烤皿內，淋上做法 (2)，鋪上做法 (4) 再均勻的淋上蛋黃，撒上起司和帕瑪森乳酪，移入烤箱以上火 220℃ 烤至呈金黃色撒上西洋香菜末。

Ingredient:

Rice 2 Cups, Western Mixed Soup Stock 2 Cups (please see p.11), Butter 4T, Crab Meat or Compressed Crab Meat 150g, Shrimp 200g, Mussel 8, Squid 1, Mushroom 12 (sliced), Finely-chopped Onion ½, Parmesan some, Egg Yolk 2, Cheese 160g, White Wine 80cc, Milk 1 Cup, Flour 4T, Finely-chopped Parsley few

Procedure:

1. Add 1 ½ T of butter in a pot to melt and add finely chopped onion (leave 1 tablespoon for later use) to sauté. Add sliced mushroom to slightly stir. Add rice to stir with low heat for about 1 minute. Add soup stock and cover with lid to cook till smoking. Then turn to low heat. For general type of pots, cook for about 25 to 30 minutes. For HotPan, turn to low heat cooking for 6 minutes upon smoking and then turn off the heat. Remove to outer cooker to simmer for 6 minutes.
2. Melt 2T of butter in a pot and add flour to stir with low heat. After stirring 8 minutes later, add milk to blend till smooth. Add 40g cheese and few pinches of salt to mix evenly.
3. Cut squid into rings. Shell the shrimps with tails left.
4. Add ½ T of butter to melt and add 1 tablespoon of finely chopped onion to sauté. Add seafood to slightly stir and drip white wine to cook till evaporated. Sprinkle some black pepper powder and salt as seasoning.
5. Add Procedure (1) into baking container and pour Procedure (2). Put Procedure (4) on top. Then evenly pour whisked egg yolks, sprinkle cheese and Parmesan, and put into oven to bake till golden with top heat 220℃ . Sprinkle finely chopped parsley as final garnish.

完美烹調寶典
Perfect Cooking Tips

- 可選用自己喜愛的海鮮。
- 可選用現切的起司，如 "EMMENTAL" 或其他自己喜愛的起司。

- You could use any kinds of seafood upon personal preference.
- You could choose freshly-cut cheese such as "EMMENTAL," or other preferred cheeses.

CC的焗飯料理中這道是最受喜愛的，有層次的口感，白醬的香濃和材料的多元性，擄獲了很多人的胃。

Among CC's risotto and gratin dishes, this one is probably one of the most favorite by others because of the rich layers of textures and taste, savory smell and creamy flavor of white sauce and diversity of food ingredients. It enchants a lot of people's stomach.

梅西那干貝蘆筍燉飯

Messina Scallop and Asparagus Risotto

材料：

義大利米 2 杯、西式高湯 1000cc（參見 p.9）、新鮮干貝 200g、蘆筍 150g（切丁）、西芹 80g（削皮切丁）、中芹 2 大匙（切末）、蒜末 1 大匙、洋蔥末 2½ 大匙、白酒 120cc、帕瑪森乳酪適量、鹽適量、奶油 ½ 大匙

做法：

1. 鍋入 1½ 大匙油待熱倒入蒜末爆炒至呈微黃色，即入洋蔥末炒香，倒入義大利米以小火炒至呈象牙白色，淋上 100cc 白酒使之蒸發。
2. 先將高湯注入做法 (1)（蓋住米的量），再煮至高湯收至快乾再倒入高湯，邊煮邊加高湯約煮 12 分鐘放入西芹丁、蘆筍丁再約煮 3 分鐘。
3. 鍋入 ⅓ 大匙油待熱倒入洋蔥爆香，放入切大丁塊的干貝，淋上白酒快炒幾下即可。
4. 做法 (3) 倒入做法 (2) 並放入奶油拌勻，以適量的鹽調味，再撒上 1½ 大匙的帕瑪森乳酪及中芹拌勻，盛入盤內撒上帕瑪森乳酪。

Ingredient:

Italian Rice 2 Cups, Western Style Soup Stock 1000cc (please see p.9), Raw Scallop 200g, Diced Asparagus 150g, Peeled and Diced Salary 80g, Finely-chopped Chinese Salary 2T, Finely-chopped Garlic 1T, Finely-chopped Onion 2 ½ T, White Wine 120cc, Parmesan few, Salt some, Butter ½ T

Procedure:

1. Add 1 ½ T of oil to heat and add finely chopped garlic to sauté till slightly yellowish. Add finely chopped onion to sauté. Add Italian rice to stir with low heat till ivory. Drip 100cc of white wine to cook till evaporated.
2. First add soup stock into Procedure (1) (till cover the rice). Then cook till soup stock almost reduced to dry and add more soup stock. Pour soup stock while continue cooking for about 12 minutes. Add diced salary and asparagus to cook for 3 minutes.
3. Add ⅓ T of oil in a pot to heat and add finely chopped onion to sauté. Add chucked scallops and drip white wine to stir quickly.
4. Add Procedure (3) into Procedure (2) and then add butter to mix well. Add few pinches of salt as seasoning. Then Sprinkle 1 ½ T of Parmesan and Chinese salary to blend evenly. Place to a plate and sprinkle more Parmesan.

完美烹調寶典
Perfect Cooking Tips

- 燉義大利米需使用木杓拌勻，才不易破壞米粒的完整性。
- 義大利米不需洗，才能吸收住高湯，如洗過水，高湯吸收不住。
- 西芹一定要削去外層較粗的纖維，才不會口感太柴；中芹亦指一般的芹菜，本書中增加芹菜為使之更增添其燉飯香。

- When cooking Italian rice, you need to use wooden spoon to stir so the rice won' t be broke or damaged.
- Italian rice needs not to rinse in advanced so that it could absorb soup stock. If you wash it with water, it won' t suck enough soup stock.
- The outer skin of rough fiber must be peeled from salary so that the texture won' t be too harden. Chinese salary sometimes refers to salary in general and it would enhance more peculiar fragment of salary. In this recipe, finely chopped salary is added to enhance savor of the risotto.

義大利的燉飯有多種口味，而 CC 最佳的首選之一就是這道，它真的非常耐吃，又帶清香的味道，很令人陶醉。

There are many flavors of Italian risotto. CC's No. 1 choice of risotto is this dish. You can never get bored of eating it while the fresh aroma is so good. This is a very enchanting dish.

瓦倫西亞燉飯

Valencia Seafood Risotto

材料：
去骨雞腿 2 隻（1 隻切 4 塊）、義大利米 2 杯、牛番茄 3 顆、四季豆 100g、紅甜椒 1 顆（切塊）、洋蔥末 3 大匙、蒜末 2 大匙、蝦 8 隻、西式綜合高湯 1000cc（參見 p.11）、白酒 150cc、番紅花 ¼ 大匙

醃料：
黑胡椒粉適量、鹽適量、香蒜粉 1 小匙、紅椒粉適量

做法：
1. 雞肉正反兩面撒上醃料放置 30 分鐘。
2. 番紅花 3 大匙放入溫熱水中浸泡約 8 分鐘，牛番茄去皮切丁狀，四季豆切斜段。
3. 雞肉入鍋內煎至金黃色取出備用。
4. 橄欖油 ½ 大匙倒入做法 (3) 的鍋內，放入蒜末爆香至微黃，再入洋蔥末爆香，倒入米炒至呈現象牙白之色，淋上白酒使之蒸發，倒入高湯蓋住米燉煮至高湯吸至快乾，再倒入高湯（邊煮邊加），燉煮約 8 分鐘後倒入番紅花和浸泡的湯汁拌勻，再放入牛番茄燉煮 3 ～ 5 分鐘，放入蝦、四季豆和紅甜椒煮至熟，以適量鹽調味，拌勻放上做法 (3) 撒上西洋香菜。

Ingredient:
Boned Drumstick 2 (Cut each one into 4 chucks), Italian Rice 2 Cups, Beef Tomato 3, String Bean 100g, Red Pepper 1 (chucked), Finely-chopped Onion 3T, Finely-chopped Garlic 2T, Shrimp 8, Western Mixed Soup Stock 1000cc (please see p.11), White Wine 150cc, Saffron ¼ T

Marinate:
Black Pepper Powder Salt, Garlic Powder 1t, Red Pepper Powder few

Procedure:
1. Sprinkle the marinate mixture onto both side of chicken and leave for 30 minutes.
2. Soak 3 tablespoons of saffron in warm water for 8 minutes. Peel and dice beef tomato. Cut string beans into segments.
3. Pan-fry the marinated chicken till golden brown. Take the pan-fried chicken out and leave aside.
4. Add ½ T of Olive oil into the pot of Procedure (3) and add finely chopped garlic to sauté till slightly yellowish. Add finely chopped onion to sauté. Add rice to stir till ivory. Drip white wine to cook till evaporated. Add soup stock to cook. While cooking, slowly add more soup stock to braise about 8 minutes. Add saffron and saffron juice to blend evenly. Add beef tomato to cook for another 3 to 5 minutes. Add shrimps, string bean and red pepper to cook till done. Add few pinches of salt as seasoning. Mix well and add Procedure (3). Sprinkle parsley as final garnish.

・番紅花浸泡的水不可丟掉，一起放入米內拌勻。

・Blend the water soaking with saffron into rice well. Do not throw away.

西班牙海鮮鍋飯是世界知名的料理，而瓦倫西亞燉飯和它有異曲同工之妙，也是西班牙著名的鍋料理之一。

The Paella is one of the most famous Spanish dishes on the world while the Valencia risotto shares similar qualities and features with it, which is another well-renowned Spanish pot dish.

法式豬排蓋飯

French Style Deep-fried Pork Chop Rice Bowl

帶有法式風味的蓋飯，在蓋飯的種類中是最令青少年喜愛的。
This rice bowl dish is very much French style and thus very popular to teenagers among so many different types of rice bowl dishes.

材料：

豬排（豬里肌肉）4 片約 600g、米 1½ 杯、西式高湯 1½ 杯（參見 p.9）、奶油 2 大匙、蘑菇 120g（切片）、洋蔥末 3 大匙、洋蔥 ½ 顆（切絲）、白蘭地 2 大匙、蒜末 ½ 大匙、牛番茄 1 顆（切片）、EMMENTAL 乳酪 4 片、麵包粉適量、蛋 2 顆、麵粉適量、西洋香菜少許、濃高湯 ½ 杯

醃料：

黑胡椒粉與鹽適量（比例為 2：1）、香蒜粉 ½ 小匙

調味料：

鹽適量

做法：

1. 豬排以肉捶棒拍打使之鬆弛，撒上醃料放置 15 分鐘。
2. 鍋倒入 1½ 大匙奶油待融化，放入洋蔥末 3 大匙炒至呈透色，放入蘑菇片炒，再倒入米（小火）炒至呈現象牙白，注入高湯放入適量的鹽拌勻，蓋上鍋蓋待冒煙改小火約煮 20 ～ 25 分鐘（使用休閒鍋待冒煙改小火煮約 6 分鐘，移外鍋燜 6 分鐘）。
3. 將做法 (1) 充分的沾上麵粉，均勻的沾裹住蛋液，平均的沾上麵包粉，放入鍋中煎至 8 分熟取出放於烤盤內。
4. 鍋入 ½ 大匙奶油待融化放入蒜末、洋蔥絲爆香炒至呈透色，淋上白蘭地倒入濃高湯約煮 6 分鐘。
5. 做法 (4) 淋於做法 (3) 鋪上乳酪入烤箱以上下火 220 ～ 230℃ 至乳酪融化取出。
6. 做法 (2) 置於盤中再鋪上做法 (5)，放上牛番茄片，撒上西洋香菜。

Ingredient:

Pork Chop (Pork Tenderloin) 4 pieces about 600g, Rice 1 ½ Cup, Western Style Soup Stock 1 ½ Cup(please see p.9), Butter 2T, Mushroom (sliced) 120g, Finely-chopped Onion 3T, Shredded Onion ½, Brandy 2T, Finely-chopped Garlic ½ T, Beef Tomato 1 (sliced), EMMENTAL Cheese 4 pieces, Bread Powder few, Egg 2, Flour some, Parsley few, Thick Soup Stock ½ Cup

Marinate:

Black Pepper Powder some, Salt some, Garlic Powder ½ t

*The proportion between black pepper powder and salt is 2：1.

Seasoning:

Salt some pinches

Procedure:

1. Slap pork chops with tenderize to make them loose. Sprinkle marinate ingredient and leave for 15 minutes.
2. Melt 1 ½ T of butter in a pot to stir 3 tablespoons of finely chopped onion till transparent. Add sliced mushroom to stir. Add rice to stir with low heat till ivory. Add soup stock with some pinches of salt to mix well. Cover with lid to cook till smoking and turn to low heat to cook 20 to 25 minutes (If using HotPan, turn to low heat to cook for 6 minutes upon smoking. Remove to outer cooker to simmer for another 6 minutes.).
3. Dip Procedure (1) fully with flour, whisked egg and bread powder by the said order. Pan-fry till almost done. Put the pan-fried pork chops in the baking tray.
4. Melt ½ T of butter in a pot and add finely chopped garlic and shredded onion to sauté till transparent. Drip Brandy and think soup stock to cook for 6 minutes.
5. Pour Procedure (4) onto Procedure (3) and top with cheese. Put into oven with both top and bottom heat of 220 to 230℃ till cheese melted. Take the baking tray out.
6. Place Procedure (2) onto a plate and top with Procedure (5) aside. Top with sliced beef tomato and sprinkle parsley.

完美烹調寶典
Perfect Cooking Tips

- 將黑胡椒粉和鹽以 2：1 比例拌勻，其味道最適中。
- 麵粉和蛋液需很平均包裹住麵包粉才能均勻沾上。
- 肉的厚度需平均，在拍打時需留意，才會使煎好的豬排軟嫩均勻。

- Blend black pepper powder and salt with the proportion of 2：1 so that for the most perfect flavor.
- Flour and whisked egg need to be very evenly attached to pork chops so that bread powder could be coated nicely.
- Be aware of the thickness of port chop during slapping so it would be cooked perfectly soft and juicy later.

佛羅倫斯鮮蝦燉飯

Florence Shrimp Risotto

材料：

蝦 300g、蛤蠣 300g、義大利米 2 杯、西式綜合高湯 1000cc（參見 p.11）、蒜末 2 大匙、紅辣椒末 1 大匙、羅勒末 1½ 大匙、奶油 1 大匙、帕瑪森乳酪適量、白酒 180cc、牛番茄 2 顆（切丁）

做法：

1. 蝦去殼留尾巴，背部劃開取腸泥。
2. 鍋入 1 大匙油待溫熱倒入蒜末爆香至微黃，再倒入紅辣椒末略炒幾下，放進義大利米以小火炒至呈現象牙白之色，淋上 120cc 白酒使之蒸發，加入高湯（第一次加高湯加至蓋過米）邊煮邊加邊攪拌，約煮 14 分鐘。
3. 鍋入 ½ 適量油待熱放入 ½ 大匙蒜末爆香，倒入蝦炒幾下煮變色取出，再倒入蛤蠣待微開淋上白酒煮至蛤蠣全開，湯汁倒入做法 (2) 取出蛤蠣肉。
4. 待做法 (2) 燉煮 14 分鐘後，放入牛番茄丁及做法 (3) 中的蝦肉與蛤蠣肉，放入奶油及鹽拌勻再撒上 2 大匙的帕瑪森乳酪拌勻盛入盤中，撒上羅勒及少許的帕瑪森乳酪。

Ingredient:

Shrimp 300g, Clam 300g, Italian Rice 2 Cups, Western Mixed Soup Stock 1000cc (please see p.11), Finely-chopped Garlic 2T, Finely-chopped Red Chili Pepper 1T, Finely-chopped Basil Leaf 1 ½ T, Butter 1T, Parmesan few, White Wine 180cc, Beef Tomato 2 (diced)

Procedure:

1. Shell the shrimps with tails left. Cut from the back of shrimp to remove the intestine mud.
2. Add 1 tablespoon of oil to a pot to heat and add finely chopped garlic to sauté till slightly yellowish. Add finely chopped red chili pepper to stir slightly. Add Italian rice to stir (low heat) till ivory. Drip 120cc of white wine to cook till evaporated. Add soup stock (the first time add till covering the rice) to cook. Constantly add soup stock while cooking for 14 minutes.
3. Add some oil in a pot to heat and add ½ T of finely chopped garlic to sauté. Add the processed shrimps to stir gently till colored. Take the stirred shrimps out and add clams to stir till slightly open. Add white wine to cook with clams till done. Add the juice to Procedure (2) and take the clam meat out.
4. Upon Procedure (2) braising for 14 minutes, add beef tomato dices and shrimps and clam meat from Procedure (3). Add butter and salt to blend evenly. Sprinkle 2T of Parmesan to mix well. Place to a plate and sprinkle basil leaf and Parmesan as final garnish.

・不吃辣可不放紅辣椒末。
・帕瑪森乳酪具有鹹度，調味時須留意。

- Finely chopped red chili pepper is optional for those who prefer not to eat spicy food.
- Parmesan is already salty. Be aware of this when add seasoning.

最令人思念的燉飯，讓人猶如在海灘邊品嘗著充滿了海味的海鮮料理。這就是它常讓人在腦海中浮現出海島渡假風的美味料理。

As one of the risottos craving for the most, this dish often makes people feel like tasting fresh and rich seafood cuisine by seashore. Truly it is a very delicious dish which often reminds people of having vocation at beach resort.

義大利海鮮墨魚燉飯

Italian Seafood and Cuttlefish Risotto

材料：

義大利米 2 杯、西式綜合高湯 1000cc（參見 p.11）、蒜末 2 大匙、紅辣椒末適量、白酒 180cc、奶油 1 大匙、牛番茄 2 顆（去皮切丁）、南瓜 150g（去皮去籽切丁）、鮮香菇 5 朵（切丁）、蘑菇 10 朵（切丁）、墨魚（或透抽）300g（切丁）、墨魚 3 隻、墨魚醬 2 大匙、帕瑪森乳酪 3 大匙、西洋香菜少許

做法：

1. 將墨魚醬放於碗內倒入 20cc 白酒拌勻（增加香味去除腥味）。
2. 鍋入 1½ 大匙油放進蒜末爆香，即入紅辣椒末略炒幾下，倒入義大利米以小火炒至呈象牙白色，淋上 130cc 的白酒改中大火使之酒精蒸發，注入高湯（第一次注入高湯加至蓋住米）煮至高湯快吸乾再加高湯，邊燉煮邊加高湯，約煮 10 分鐘放入做法 (1) 拌勻再倒入南瓜丁及菇類和墨魚丁，再約煮 4 分鐘倒入牛番茄拌勻，並以適量的黑胡椒粉、鹽調味再撒上帕瑪森乳酪拌勻，盛入器皿內。
3. 墨魚入鍋煎至快熟淋上 30cc 白酒後再取出。
4. 將做法 (3) 擺於做法 (2) 上撒上帕瑪森乳酪及西洋香菜。

Ingredient:

Italian Rice 2 Cup, Western Mixed Soup Stock 1000cc (Please see p.11), Finely-chopped Garlic 2T, Finely-chopped Red Chili Pepper some, White Wine 180cc, Butter 1T, Beef Tomato 2 (peeled and diced), Pumpkin 150g (peeled, seeded and diced), Fresh Shitake Mushroom 5 (diced), Mushroom 10 (diced), Cuttlefish(Squid) 300g (diced), Cuttlefish 3, Cuttlefish Paste 2T, Parmesan 3T, Parsley few

Procedure:

1. Add cuttlefish paste in a bowl and pour 200cc of white wine to mix well (increase savor and erase the fishy smell).
2. Add 1 ½ T of oil in a pot to sauté finely chopped garlic. Add finely chopped red chili pepper to stir slightly. Add Italian rice to stir with low heat till ivory. Pour 130cc of white wine (turn to medium to high heat) to cook till evaporated. Add soup stock (first time the soup stock covers up the rice) to cook till soup stock reduced and add more. Add soup stock while cooking for about 10 minutes. Then add Procedure (1) to blend evenly. Add pumpkin dices, all kinds of mushrooms, and cuttlefish dices to cook. 4 minutes later add beef tomato to mix well. Add some black pepper powder and salt as seasoning. Sprinkle Parmesan to mix evenly. Place to a container.
3. Pan-fry cuttlefish till almost done. Sprinkle 30cc of white wine and take out.
4. Place Procedure (3) on top of Procedure (2) and sprinkle Parmesan and parsley as final garnish.

完美烹調寶典
Perfect Cooking Tips

- 市售義大利米烹煮時間約 12 ～ 15 分鐘，視不同廠牌。
- 義大利人約吃 8 ～ 9 分熟較有口感的義大利燉飯，但國人較不易接受，通常大部分都只能接受全熟的米。
- 義大利米在台北 SOGO 百貨（復興店）、微風百貨等超市均有販售。

- The average cooking time of Italian rice available at supermarket is between 12 to 15 minutes, depending on different brands.
- The texture of risotto or rice cooking for Italian people is al dante, which refers to almost cooked. Yet most Taiwanese people could not accept this texture and prefer well-cooked rice.
- Italian rice is available at supermarket at Taipei SOGO Department Store (Fuxing Branch), Breeze Department Store and so on.

義大利燉飯料理中以墨魚風味最受歡迎，這道墨魚燉飯有別於一般的做法，增加了一些蔬菜，使得視覺上更加亮眼，味覺更加爽口。

Among so many Italian risotto dishes, cuttlefish flavor is probably one of the most popular dishes. In this recipe which is quite different from the average way, more vegetables are added to create more eye-catchy visual effects as well as much fresh texture.

充滿南洋風味的

東南亞料理

東南亞風味滷肉飯

Southeast Asian Style Braised Pork Rice Bowl

材料：

五花肉 1kg、蒜頭 16 粒（去皮）、香菜頭 5 個、香茅 2 枝（切段）、南薑 3 片、檸檬葉 6 片（撕開）、水 1½ 杯、八角 2 粒、肉桂 1 支、白飯適量

調味料：

椰子糖（棕櫚糖）2 大匙、老抽 4 大匙、醬油 6 大匙、魚露 1 大匙

做法：

1. 五花肉切成厚度約 2cm，長約 8cm，入鍋煎至呈金黃色取出。
2. 蒜頭放入做法 (1) 的鍋內炒約 1 分鐘後，放入香菜頭，再放進其他材料和調味料燉滷約 1 小時（使用壓力鍋，不需放水，老抽 2 大匙、醬油 1½ 大匙、魚露 ½ 大匙，待壓力鍋上升二條紅線改小火約 20 分鐘）。
3. 白飯盛入碗內，放上適量的做法 (2)，擺上香菜葉即可。

Ingredient:

Pork Belly 1kg, Garlic (peeled) 16, Coriander Root 5, Lemon Grass 2 (segmented), Galangal 3 slices, Lemon Leaf 6 (torn), Water 1 ½ Cups, Aniseed 2, Cinnamon 1, Steam Rice some

Seasoning:

Coconut Sugar (Palm Sugar)2T, Dark Soy Bean Sauce 4T, Soy Bean Sauce 6T, Fish Sauce 1T

Procedure:

1. Cut pork belly into strips of 2cm thick and 8cm long. Pan-fry till golden brown. Take the pan-fried pork belly strips out.
2. Add garlic in the pot of Procedure (1) to sauté for 1 minute. Add coriander roots, Procedure (1) and other ingredients as well as seasoning to braise for 1 hour (If using a pressure cooker, there is no need to add water. Add dark soy bean sauce 2T, soy bean sauce 1 ½ T and fish sauce ½ T instead. Turn to low heat to cook about 20 minutes when it rises to two red strips.).
3. Place rice in a container and put Procedure (2) with sprinkling coriander leafs to serve.

完美烹調寶典 Perfect Cooking Tips

- 烹調時可加入約 1 小匙的肉桂粉亦有不同的風味。
- 醬油量僅參考，視各廠牌之鹹度作調整。
- 如無椰子糖可以冰糖代替。

- Cinnamon powder is about 1t and it will bring out different flavor.
- The proportion of soy bean sauce listed here is only for reference. It is subject to the salinity of each brand.
- If coconut sugar is not available, you could substitute with crystal sugar.

第一次吃到這種口味的滷肉飯，就迷上它的獨特香味，
CC再次發揮「滷」功請當地朋友問到其做法。

When having this kind of braised pork rice bowl for the first time, CC immediately felt for the unique savor. Hence once again CC persuaded local friends to ask for the recipe and cooking tips.

馬來風味海鮮炊飯

Malayan Style Seafood Casserole Rice

材料：
鮭魚 200g、米 2 杯、泰式綜合高湯 2 杯（參見p.9）、蝦 200g、透抽 1 隻、香茅 2 枝、檸檬葉 3 片、紅辣椒 1 枝、香菜末適量、蒜末 1 大匙、紅蔥頭末 1 大匙、南薑末 ½ 大匙、薑黃粉（鬱金香粉） ½ 大匙

調味料：
鹽適量

做法：
1. 鮭魚撒上適量的鹽放置 15 分鐘後，放入鍋內煎至熟取出，待涼剝成小碎塊；蝦去殼留尾巴取出腸泥；香茅切段；檸檬葉切絲；紅辣椒切小圈狀。
2. 鍋內倒入 1 大匙油待熱，放入紅蔥頭爆香，再倒入蒜末炒至微黃，即入南薑末炒香，放入米炒以小火炒至呈象牙白，倒入薑黃粉，拌勻呈黃色狀，注入高湯放入適量的鹽拌勻，鋪上香茅，蓋鍋蓋以中小火煮至冒煙改小火，約 22 分鐘放入蝦及整隻的透抽，再煮 8 分鐘即可（使用休閒鍋，待冒煙改小火煮約 4 分鐘，再放入蝦及透抽煮至冒煙移入外鍋燜 8 分鐘）。
3. 打開鍋蓋撒上鮭魚肉、檸檬葉、紅辣椒圈及香菜末。

Ingredient:
Salmon Filet 200g, Rice 2 Cups, Thai Style Mixed Soup Stock (please see p.9) 2 Cups, Shrimp 200g, Squid 1, Lemon Grass 2, Lemon Leaf 3 pieces, Red Chili Pepper 1, Finely-chopped Coriander few, Finely-chopped Garlic 1T, Finely-chopped Shallot 1T, Finely-chopped Galangal ½ T, Turmeric Powder ⅔ T

Seasoning:
Salt few pinches

Procedure:
1. Sprinkle few pinches of salt onto salmon filet and leave for 15 minutes. Pan-fry till well cooked. Wait to cool down and then torn into small pieces. Shell the shrimps with tails left and remove the intestine mud. Cut lemon grass into segments. Shred lemon leafs. Cut red chili peppers into small rings.
2. Add 1T of oil to heat and add shallot to sauté. Then add finely chopped garlic to stir till slightly yellowish. Add finely chopped galangal to sauté. Add rice to stir with low heat till ivory white. Add turmeric powder and blend evenly till turning yellow. Pour soup stock with few pinches of salt and mix well. Top with lemon grass and cover with lid to cook with medium to low heat till smoking. Then turn to low heat and around 22 minutes later add shrimps and the whole piece of squid. Cook for another 8 minutes. (If using HotPan, turn to low heat upon smoking to cook 4 minutes. Then add shrimp and squid to cook till smoking. Remove to outer cooker to simmer 8 minutes.)
3. Open the lid, put squid rings back to pot and sprinkle salmon, lemon leafs, red chili pepper rings and finely chopped coriander.

完美烹調寶典
Perfect Cooking Tips

- 鮭魚有售薄鹽鮭魚，就不需撒鹽。
- 如買不到南薑，用嫩薑代替。
- 沒有檸檬葉以蔥綠絲代替。

- If you got thin-salt preserved salmon filet, there is no need to sprinkle salt.
- If galangal is not available, you could substitute with fresh ginger.
- If lemon leaf is not available, you could substitute with the green stalk part of shredded green onion.

常常被要求再做一次的炊飯料理非常討喜，平常吃或過年節慶時超級受歡迎！

This casserole rice dish has remained very popular upon constant requests. Either for ordinary days or holiday feast, it is such high demanding dish!

印度優格雞肉飯

India Chicken with Yogurt Sauce on Rice

很多人會因為印度料理中的香料味道過重，而放棄其美味的精典印度菜，CC將其香料比例降低而不失印度菜的味道，受到很多學生們的喜歡。

Many locals don't eat very delicious and classy Indian cuisine just because the heavy use of spices. CC adjusts the proportion of spices without compromising the authentic flavor of Indian dish. Therefore such dishes are welcomed by many students.

材料：

去骨雞腿 3 隻、葡萄乾適量、原味優格（無糖）160cc、醬油 1 大匙、酒 1 大匙、香菜適量、蒜末 1½ 大匙、嫩薑末 ½ 大匙、紅辣椒末 1 大匙、咖哩粉 ½ 大匙、小茴香 1 小匙、薑黃粉 1 大匙、米 2 杯、泰式綜合高湯或西式綜合高湯 2 杯（參見 p.9/p.11）、蝦米 1 大匙、胡椒粉少許

醃料：

薑黃粉 ⅓ 小匙、咖哩粉 ⅓ 大匙、胡椒粉 1 小匙、醬油 1⅓ 大匙、鹽 1 小匙、香菜梗末 1 大匙、砂糖 ½ 小匙拌勻

做法：

1. 雞腿肉切塊拌入醃料拌勻置 30 分鐘，蝦米洗淨浸泡米酒約 6 分鐘取出剁碎。
2. 鍋入 ½ 大匙油待熱放入蒜末爆香，即入蝦米炒香倒入米小火炒約 1 分鐘，放入 1 大匙薑黃粉炒勻，注入高湯將飯煮熟。
3. 鍋入 ½ 大匙油待熱入做法 (1) 的雞肉，皮先朝鍋底煎至金黃色取出。
4. 蒜末放入做法 (3) 的鍋內爆香即入薑末及香菜梗末爆香，倒入小茴香改小火炒，並放入咖哩粉 ½ 大匙略炒幾下，放入優格 ⅓ 大匙薑黃粉、胡椒粉、醬油 1 匙及少許的鹽拌勻，倒入做法 (3) 煮約 6 分鐘，最後撒上香菜末及紅辣椒末，盛入器皿內。
5. 將煮好的飯盛入器皿內撒上葡萄乾，和做法 (4) 一起食用。

Ingredient:

Boned Drumstick 3, Ram Raisin some, Original Flavor Yogurt (sugar free) 160cc, Soy Bean Sauce 1T, Wine 1T, Coriander some, Finely-chopped Garlic 1 ½ T, Finely-chopped Fresh Ginger ½ T, Finely-chopped Red Chili Pepper 1T, Curry Powder ½ T, Cumin 1t, Turmeric Powder 1T, Rice 2 Cups, Thai Style Mixed Soup Stock or Western Mixed Soup Stocl 2 Cups (please see p.9/p.11), Shelled Dried Small Shrimp 1T, Pepper Powder few

Marinate:

Turmeric Powder ⅓ t, Curry Powder ⅓ T, Pepper Powder 1t, Soy Bean Sauce 1 ⅓ T, Salt 1t, Finely-chopped Coriander Stalk 1T, Fine Sugar ½ t. Blend all ingredients well.

Procedure:

1. Cut drumsticks in chucks and blend with marinate evenly. Leave for 30 minutes. Rinse dried shelled small shrimps and soak in rice wine for 6 minutes. Take the soaked shrimps out and cut up.
2. Add ½ T of oil in a pot to heat and add finely chopped garlic to sauté. Add dried shelled small shrimps to stir till savor and then add rice to stir with low heat for 1 minute. Add 1 tablespoon of turmeric powder to stir evenly. Pour soup stock to cook the rice till done.
3. Add ½ T of oil in a pot to heat and add chicken from Procedure (1) to pan-fry till golden brown. Have the skin side facing down against pan first. Take the pan-fried chicken out.
4. Add finely chopped garlic in the pot of Procedure (3) to sauté and add finely chopped ginger and coriander stalk to sauté. Add cumin and turn to low heat. Add ½ T of curry powder to slightly stir. Add yogurt ⅓ T, turmeric powder, pepper powder, 1 spoon of soy bean sauce and few pinches of salt to mix well. Add Procedure (3) to cook for 6 minutes. Lastly sprinkle finely chopped coriander and red chili pepper. Place to a container.
5. Add cooked rice in a container and sprinkle some ram raisin. Serve with Procedure (4).

完美烹調寶典
Perfect Cooking Tips

- 亦也搭配法國麵包或吐司。
- 市售優格酸度不夠可放少許檸檬汁。

- This recipe could also serve with baguette or toast.
- The yogurt available at supermarket is not sour enough, you could add some lemon juice.

泰式烤牛排飯

Thai Style Roasted Steak Fried Rice

材料：
沙朗或菲力牛排 2 塊、白飯 2 碗、蛋 2 顆（拌成蛋液）、蒜末 2 大匙、蔥末 2 大匙、檸檬 1 個、熟筍丁 ½ 杯

醃料：
檸檬葉 2 片（切絲）、魚露 ½ 大匙、醬油 1 大匙、白砂糖 1 小匙、蒜末 ½ 大匙、香菜末 ½ 大匙、檸檬汁 ½ 大匙、黑胡椒粉 1 小匙拌勻

調味料：
白胡椒粉少許、魚露 1½ 大匙、醬油 1½ 大匙

配菜：
小黃瓜 2 條（切片）、紅蔥頭 6 個（切片）、紅辣椒 1 枝（切圈狀）、白醋 5 大匙、白砂糖 4 大匙、鹽 1 大匙、洋蔥 ¼ 顆（切絲）；白醋、白砂糖、鹽放入鍋內煮勻待涼，後放入其他材料拌勻放置 1 小時。

做法：
1. 牛排放入醃料拌勻置 1 小時後，入鍋煎至自己喜愛的熟度取出。
2. 鍋內放入 2 大匙油待熱放入 1 大匙蒜末及 1 大匙蔥末爆香，即入筍丁略炒幾下，倒入蛋液拌勻幾下，放入白飯倒入調味料拌炒勻，最後再撒上 1 大匙蒜末及蔥花拌勻即可，盛盤鋪上牛排，放上 ½ 顆檸檬，附上配菜。

Ingredient:
Sirloin or Fillet Mignon 2 pieces, Steam Rice 2 Bowls, Whisked Egg 2, Finely-chopped Garlic 2T, Finely-chopped Green Onion 2T, Lemon 1, Cooked Bamboo Shoot Dices ½ Cup

Marinate:
Lemon Leaf 2 pieces (shredded), Fish Sauce ½ T, Soy Bean Sauce 1T, Fine White Sugar 1t, Finely-chopped Garlic ½ T, Finely-chopped Coriander ½ T, Lemon Juice ½ T, Black Pepper Powder 1t. Blend all ingredients well.

Seasoning:
White Pepper Powder few, Fish Sauce 1 ½ T, Soy Bean Sauce 1 ½ T

Procedure:
Cucumber 2 (sliced), Shallot 6 (sliced), Red Chili Pepper (cut into rings), White Vinegar 5T, Fine White Sugar 4T, Salt 1T, Shredded Onion ¼
Cook white vinegar, fine white sugar and salt in a pot till melting evenly. Wait for cool down. Then add other ingredients to blend well. Leave for 1 hour.

Procedure:
1. Blend steaks with marinate well and leave for 1 hour. Pan-fry as individual preference. Take the pan-fried steaks out.
2. Add 2T of oil in a pot to heat and add 1T of finely chopped garlic and green onion respectively to sauté. Add cooked bamboo shoot dices to stir slightly. Add whisked egg to stir gently. Add steam rice and seasoning to stir and blend evenly. Lastly sprinkle 1T of finely chopped garlic and green onion to mix well. Place to a plate and top with steak. Serve with side dish and half lemon.
Side Dish

・可改用去骨雞腿肉或豬排。

・You could substitute with boned drumsticks or pork chops.

具有飽足感的一道料理年輕人特別喜歡，使用泰國香料醃過的牛排，更具有清香味不油膩。

Such big meal is so fond by young people. Steak marinated with Thai spices has fresh smell without greasy taste.

越南順化燉肉

Vietnam Huế Beef Stew

材料：

牛肋條（牛腩）1kg、洋蔥 1 顆（切末）、紅蔥頭末 3 大匙、蒜末 2 大匙、香茅 3 枝、紅蘿蔔 1 條、椰漿 1 罐、白飯適量、咖哩粉 3 大匙、西式牛高湯 3 杯（參見 p.9）、檸檬 2 顆

醃料：

咖哩粉 1½ 大匙、蒜末 1 大匙、紅辣椒粉適量、魚露 1 大匙、白砂糖 ⅓ 大匙

調味料：

魚露 4 大匙、白砂糖 ½ 大匙、少許鹽

做法：

1. 牛肋條切長約 5cm，拌入醃料拌勻置 30 分鐘後放入鍋內，煎至呈焦黃色取出備用。
2. 1 大匙油倒入做法 (1) 的鍋內，放入洋蔥炒至呈透色狀倒入蒜末爆香，再入紅蔥頭末炒香，改小火倒入咖哩粉炒至香味溢出，注入高湯放入紅蘿蔔（切滾刀塊狀）及做法 (1) 和香茅（切段）燉煮約 1 小時（使用壓力鍋高湯只需放 ¼ 杯，待上升二條紅線改小火 20 分鐘）後倒入椰漿煮滾放入調味料拌勻。
3. 將做法 (2) 盛入器皿內再搭配白飯，附上檸檬食用時淋於燉肉內。

Ingredient:

Beef Rib Finger (Beef Brisket) 1kg, Finely-chopped Onion 1, Finely-chopped Shallot 3T, Finely-chopped Garlic 2T, Lemon Grass 3, Carrot 1, Coconut Milk 1 can, Steam Rice some, Curry Powder 3T, Western Style Beef Soup Stock 3 Cups (please see p.9), Lemon 2

Marinate:

Curry Powder 1 ½ T, Finely-chopped Garlic 1T, Red Chili Pepper Powder few, Fish Sauce 1T, Fine White Sugar ⅓ T

Seasoning:

Fish Sauce 4T, Fine White Sugar ½ T, Salt few

Procedure:

1. Cut beef rib finger into strips of 5cm long. Blend with marinate well and leave for30 minutes. Pan-fry till golden brown and take the pan-fried beef out. Leave aside.
2. Add 1T of oil in the pot of Procedure (1). Add onion to stir till transparent. Add soup stock and add carrot that is cut into round chucks, Procedure (1), and segmented lemon grass to braise for about 1 hour. (If using a pressure cooker, the soup stock only requires ¼ cup. Turn to low heat to cook 20 minutes when it rises to two red strips.) Add coconut milk to boil and then add seasoning to blend well.
3. Place Procedure (2) in a container and add steam rice. Serve with lemon to drip fresh lemon juice on top of braised beef.

完美烹調寶典
Perfect Cooking Tips

・也可使用豬梅花肉或牛腱肉。

・You could also use pork shoulder or beef shank.

越南以前是法國的殖民地，所以這道料理也可以用烤得外酥內軟的法國麵包做搭配。

Vietnam used to be colonial to France. Therefore this dish could also serve with baguette that is baked crispy outside and soft inside.

南洋辣炒雞丁飯

Southeast Asian Style Spicy Chicken Dices on Rice

材料：
去皮去骨雞腿肉 3 隻、蒜末 2 大匙、薑末 ½ 大匙、香菜末 1 大匙、紅辣椒末適量、九層塔適量、白飯適量、紅辣椒 1 枝（切段）

Ingredient:
Skinned and Boned Drumstick 3, Finely-chopped Garlic 2T, Finely-chopped Ginger ½ T, Finely-chopped Coriander 1T, Finely-chopped Red Chili Pepper few, Basil Leaf few, Steam Rice some, Red Chili Pepper 1 (segmented)

調味料：
魚露 1 大匙、老抽 1 大匙、醬油 1 ½ 大匙、泰式綜合高湯 ¼ 杯（參見 p.9）、辣椒粉適量、糖 1 小匙

Seasoning:
Fish Sauce 1T, Dark Soy Bean Sauce 1T, Soy Bean Sauce 1½T, Thai Style Mixed Soup Stock ¼ Cup(please see p.9), Chili Powder few, Sugar 1t

醃料：
檸檬葉末 1 大匙、蒜末 1 大匙、醬油 1 大匙、白胡椒粉少許、魚露 ⅓ 匙、白砂糖 1 小匙拌勻

Marinate:
Finely-chopped Lemon Leaf 1T, Finely-chopped Garlic 1T, Soy Bean Sauce 1T, White Pepper Powder few, Fish Sauce ⅓ T, Fine White Sugar 1t. Blend all ingredients evenly.

做法：
1. 雞腿肉切塊狀拌入醃料，拌勻靜置 15 分鐘後，倒入鍋內（皮先朝鍋底）煎至金黃色取出。
2. 鍋內放入 1 大匙油待熱，放入蒜末爆香再入薑末、香菜末炒香即入紅辣椒末，並倒入做法(1)拌炒勻。
3. 調味料倒入做法(2)內炒至汁收乾，最後撒上九層塔和紅辣椒段。
4. 將白飯倒入盤內舖上做法(3)。

Procedure:
1. Chop drumsticks into chucks and blend with marinate well. Leave for 15 minutes. Pan-fry till golden brown (first put the skin side facing down the pan). Take the pan-fried drumstick meats out.
2. Add 1T of oil in a pot to heat. Add finely-chopped garlic to sauté. Then add finely-chopped ginger and coriander to sauté. Add finely-chopped red chili pepper and Procedure (1) to blend well.
3. Add seasoning into Procedure (2) to stir till liquid reduced. Lastly sprinkle basil leafs and segmented red chili pepper.
4. Add steam rice in a plate and top with Procedure (3).

完美烹調寶典
Perfect Cooking Tips

- 亦可使用雞胸肉或豬小里肌肉。
- 老抽為顏色深的醬油。

- You could also use chicken breast or pork tenderloin in this recipe.
- Dark Soy Bean Sauce: dark colored soy bean sauce.

如果胃口不好，這料理一定會讓你食慾大開。
No bon appetite? This dish surely will bring you a “yes” to this question.

馬來西亞海鮮燴飯

Malaysian Style Seafood Gravy over Rice

材料：

蝦 500g、透抽 400g、牛番茄 2 顆、蒜末 2 大匙、香茅末 ½ 匙、南薑末 ½ 大匙、香菜梗末 2 大匙、太白粉 2 大匙、泰式綜合高湯 3 杯（參見 p.9）、紅辣椒末適量、米 2 杯、酒 1 匙、香菜葉適量

調味料：

辣豆瓣醬 ½ 大匙、番茄醬 2 大匙、白砂糖 1 小匙、老抽 ½ 大匙、醬油 1½ 大匙

做法：

1. 香料飯：鍋內放入 ½ 大匙油待熱後放入蒜末爆香，再入香菜梗末 1 大匙炒香，倒入米略炒 1 分鐘，注入高湯 2 杯，放入適量的鹽拌勻，鋪上香茅切段，蓋上鍋蓋待冒煙改小火煮約 25 ～ 30 分鐘。（使用休閒鍋，待冒煙改小火煮 6 分鐘，移外鍋燜 6 分鐘。）
2. 蝦去殼留尾巴，由背部劃開取出腸泥；透抽切長條狀；牛番茄去皮切丁狀。
3. 鍋放入 1 大匙油待熱放入蒜末爆香，再入南薑末、香菜梗末 1 大匙炒香，倒入辣豆瓣醬，淋上酒炒勻，再放入番茄丁及番茄醬，注入 1 杯高湯，倒入白砂糖、老抽、醬油煮沸，放入蝦及透抽煮至海鮮熟，以太白粉勾芡拌勻。
4. 香料飯盛入器皿內，淋上做法 (3) 撒上香菜葉即可。

Ingredient:

Shrimp 500g, Squid 400g, Beef Tomato 2, Finely-chopped Garlic 2T, Finely-chopped Lemon Grass ½ T, Finely-chopped Galangal ½ T, Finely-chopped Coriander Stalk 2T, Potato Starch 2T, Thai Style Mixed Soup Stock 3 Cups (please see p.9), Finely-chopped Red Chili Pepper some, Rice 2 Cups, Wine 1T, Coriander Leaf few

Seasoning:

Spicy Bean Paste ½ T, Ketchup 2T, Fine White Sugar 1t, Dark Soy Bean Sauce ½ T, Soy Bean Sauce 1 ½ T

Procedure:

1. Spice Rice:Add ½ T of oil in a pot to heat and add finely chopped garlic to sauté. Add 1 tablespoon of finely chopped coriander root to sauté. Add rice to slightly stir for 1 minute, and then pour 2 cups of soup stock. Add few pinches of salt to mix evenly. Top with segmented lemon grass. Cover with lid to cook till smoking and turn to low heat to cook about 25 to 30 minutes. (If using HotPan, turn to low heat to cook for 6 minutes upon smoking. Then remove to outer cooker to simmer for another 6 minutes.)
2. Shell the shrimps with tails left. Cut from the back of shrimp to remove the intestine mud. Cut squid into long strips. Peel the tomato and cut in dices.
3. Add 1T of oil in a pot to heat and add finely chopped garlic to sauté. Then add finely chopped galangal and 1 tablespoon of finely chopped coriander root to sauté. Add spicy bean paste and drip wine to stir evenly. Add tomato dices and ketchup with 1 cup of soup stock. Then add fine white sugar, dark soy bean sauce, soy bean sauce to boil. Add shrimps and squid to cook till done. Thicken with potato starch water.
4. Place spice rice in a container and pour Procedure (3) with sprinkling coriander leafs to serve.

完美烹調寶典
Perfect Cooking Tips

・可選用自己喜愛的海鮮。
・不吃辣者可改用豆瓣醬。

・You could use any kinds of seafood you prefer.
・For those who don' t eat spicy food, use bean paste instead.

偶爾來上一盤南洋風味的燴飯，會讓人的胃口及味蕾有著清新感和舒暢感。

Occasionally having a plate of gravy over rice with Southeast Asian flavor would bring such freshness and comfort to one's appetite as well as taste bud.

新加坡海南雞飯

Singapore Hainan Chicken Rice

材料：

仿土雞或土雞 1 隻、米 2 杯、蒜末 2 大匙、香茅 1 隻、蔥 4 枝（拍扁）、鹽適量

配菜：

紅蘿蔔去皮切丁 100g、白蘿蔔去皮切丁 200g、小黃瓜切丁 100g、拌入 1 大匙鹽，置入 30 分鐘後瀝乾水分，倒入白砂糖 4 大匙、白醋 6 大匙、月桂葉 1 片，拌勻醃 3 小時（放冰箱冷藏）

沾醬：

醬油 3 大匙、白砂糖 1 小匙、檸檬汁 ½ 大匙、嫩薑末 ½ 大匙、紅辣椒末適量、辣豆瓣醬 1 大匙、煮雞的高湯 2 大匙拌勻

做法：

1. 土雞抹上適量的鹽和蔥放置 1 小時。
2. 雞放入鍋內注入水至蓋住雞肉，蓋上鍋蓋煮滾後改小火煮約 5 分鐘，熄火燜 20 分鐘（不可打開鍋蓋）後打開鍋蓋取出雞肉，待涼剁塊狀。
3. 鍋內倒入 1 大匙油待熱放入蒜末爆香至微黃，倒入米炒約 90 秒，再入做法 (2) 煮雞的高湯 2 杯，放入適量的鹽拌勻，香茅切段鋪上蓋上鍋蓋待冒煙，改小火煮 25 ~ 30 分鐘，使用休閒鍋煮 6 分鐘移外鍋燜 6 分鐘。
4. 適量剁好的雞肉至於盤內，放上煮好的飯，附上配菜並搭配一小碟沾醬。

Ingredient:

Semi-farming Chicken or Homebrew Chicken 1, Rice 2 Cups, Finely-chopped Garlic 2T, Lemon Grass 1, Leek 4 (crush flat), Salt few

Side Dish

Peeled Carrot Dices 100g, Peeled Daikon Radish Dices 200g, and Cucumber Dices 100g. All blend with 1 tablespoon of salt and leave for 30 minutes. Drain off juice. Add fine white sugar 4T, white vinegar 6T, and 1 piece of bally leaf. Blend evenly to marinate for 3 hours (store cold in refrigerator).

Dip Sauce

Soy Bean Sauce 3T, Fine White Sugar 1t, Lemon Juice ½T, Finely-chopped Fresh Ginger ½ T, Finely-chopped Red Chili Pepper few, Spicy Bean Paste 1T, Soup Stock from Braising the Chicken 2T. Blend all ingredients well.

Procedure:

1. Wipe few pinches of salt onto chicken with leek. Leave for 1 hour.
2. Add the salt-wiped chicken in a pot and add water till covering the chicken. Cover with lid to boil and turn to low heat to cook for 5 minutes. Turn off the heat and simmer for 20 minutes (do not open the lid), and open the lid to take the chicken out. Wait for cooling down and cut into chucks.
3. Add 1T of oil in a pot to heat and add finely chopped garlic to sauté till slightly yellowish. Add rice to cook around 90 seconds and then add 2 cups of the chicken soup stock from Procedure (2) with few pinches of salt to blend evenly. Put segmented lemon grass on top and cover with lid to cook till smoking. Turn to low heat to cook another 25 to 30 minutes. If using HotPan, cook 6 minutes and then remove to outer cooker to simmer for 6 minutes.
4. Take some chucked chicken to a plate. Add cooked rice. Serve with side dish and a small dish of dip sauce.

CC幾年前為了它飛至新加坡尋覓CC理想中的海南雞肉飯，現在CC將心得與大家分享。

Few years ago, CC flight to Singapore in order to find the Hainan chicken rice that CC is so fond of. Now CC shares the recipe here with all readers.

泰式海鮮什錦炒飯

Thai Style Assorted Seafood Fried Rice

材料：

蝦 150g、透抽 100g、雞胸肉 100g、蛋 2 顆（拌成蛋液）、熟白飯 2 碗、洋蔥末 3 大匙、紅蔥頭末 1 大匙、蔥末 1 大匙、蒜末 1½ 大匙、香菜梗末 1 大匙、檸檬 ½ 顆、牛番茄 ½ 顆、大黃瓜片適量

調味料：

魚露 2 大匙、醬油 1 大匙、白砂糖 ⅓ 大匙、白胡椒粉少許

醃料：

蒜末 ¼ 大匙、香菜末 ¼ 大匙、白胡椒粉少許、魚露 ¼ 大匙、糖少許

做法：

1. 蝦去殼留尾巴去腸泥，透抽切長條狀，雞胸肉切片狀，拌入醃料拌勻置 15 分鐘。
2. 鍋內倒入 2 大匙油待熱，倒入蛋液炒散至 5 分熟取出備用，再入做法 (1) 的雞肉炒至肉呈白色取出。
3. 蒜末、紅蔥頭末倒入做法 (2) 的鍋內爆香放入洋蔥末，再放入蔥花及香菜末炒香，倒入海鮮略炒幾下即入白飯，倒入調味料拌炒再放入做法 (2) 炒勻即可盛盤。
4. 大黃瓜片，牛番茄切節狀，檸檬對切置盤邊。

Ingredient:

Shrimp 150g, Squid 100g, Chicken Breast 100g, Whisked Egg 2, Steamed Rice 2 Bowls, Finely-chopped Onion 3T, Finely-chopped Shallot 1T, Finely-chopped Green Onion 1T, Finely-chopped Garlic 1 ½ T, Finely-chopped Coriander Stalk 1T, Lemon ½, Tomato ½, Cucumber Slices few

Seasoning:

Fish Sauce 2T, Soy Bean Sauce 1T, Fine White Sugar ⅓ T, White Pepper Powder few

Marinate:

Finely-chopped Garlic ¼ T, Finely-chopped Coriander ¼, White Pepper Powder few, Fish Sauce ¼ T, Sugar few

Procedure:

1. Shell the shrimps with tails left and remove the intestine mud. Cut squid into long strips. Slice the chicken breast. Blend with marinate well and leave for 15 minutes.
2. Add 2T of oil in a pot to heat and add whisked egg to stir till medium rare. Take the scrambled egg out and leave aside. Add Procedure (1) Chickens in this pot to stir till meat turning white.
3. Add finely chopped garlic and shallot into the pot of Procedure (2) to sauté. Then add finely chopped onion, green onion and coriander to sauté. Add seafood to stir slightly and immediately add steamed rice. Add seasoning to blend and add Procedure (2) to stir and mix well. Place it to a plate.
4. Put cucumber slices, chucked beef tomato and halved lemon around the plate.

完美烹調寶典
Perfect Cooking Tips

- 食用時，淋上檸檬汁及泰式風味肉丸粥（參見 p.68）中的醬料會更美味。

- Serving with lemon juice or the sauce of Thai style meat ball congee (please see p.68) will extra delicious.

CC最喜歡在曼谷的考桑路，在任何一家的餐廳點炒飯既美味又便宜，而這道是CC必點的。

What CC likes the most about Bangkok is the fried rice at any restaurants on Khao San Road, which is always tasty and cheap. This dish in particular is the must-order one by CC..

泰國風味肉丸粥

Thai Style Meat Ball Congee

材料：

雞胸肉末（雞絞肉）300g、花枝漿 150g、米 1½、泰式綜合高湯 2000cc（參見 p.9）、香菜末、蔥末、嫩薑絲各適量、紅蔥頭酥適量

調味料 A：

香菜末 1½ 大匙、魚露 ⅓ 大匙、白胡椒粉少許、鹽少許、蛋 1 顆、玉米粉 1½ 大匙、荸薺末 3 大匙（擠掉水份）、白砂糖 1 小匙

調味料 B：

魚露 3 大匙、白砂糖 ½ 大匙、白胡椒粉少許、鹽適量

做法：

1. 雞胸肉末放入調理盆內，倒入花枝漿及調味料 A 拌勻，做成圓球狀。
2. 高湯注入鍋內放入做法 (1) 以小火煮至熟取出備用；米倒入鍋內煮成粥，再將做好的肉丸倒入，放進調味料 B 拌勻，盛入器皿內，撒上香菜末、蔥末、薑絲、紅蔥頭酥。

Ingredient:

Finely-chopped Chicken Breast (Minced Chicken) 300g, Squid Paste 150g, Rice 1 ½ Cup, Thai Style Mixed Soup Stock 2000cc(Please see p.9), Finely-chopped Coriander, Finely-chopped Green Onion and Shredded Ginger few each, Deep-fried Shallot some

Seasoning: A

Finely-Chopped Coriander 1 ½ T, Fish Sauce ⅓ T, White Pepper Powder few, Salt few, Egg 1, Corn Starch 1 ½ T, Finely-chopped Water Chestnut 3T (get rid of juice), Fine White Sugar 1t

Seasoning: B

Fish Sauce 3T, Fine White Sugar ½ T, White Pepper Powder few, Salt some

Procedure:

1. Add finely chopped chicken breast into a basin. Add squid paste and seasoning A to blend well. Shape the mixture into balls.
2. Add soup stock in a pot and boil Procedure (1) with low heat till cooked. Take the cooked meat balls out. Set aside; Add rice in a pot to cook into congee. Add the cooked meat balls and seasoning B to mix evenly. Place in a container and sprinkle finely chopped coriander, green onion, shredded ginger and deep fried shallot.

完美烹調寶典
Perfect Cooking Tips

- 這樣更好吃！做一小碗醬料淋在粥上拌勻。醬料：紅辣椒切圓圈狀 2 大匙、魚露 4 大匙、白醋或檸檬汁 1/4 杯拌勻
- 花枝漿有鹹度，只需放入雞肉部份的鹹度。
- 可使用豬絞肉。

- This Way, Taste Better! Make a small bowl of sauce dripping on the congee and mix evenly. Sauce: Red Chili Pepper (cut in rings) 2T, Fish Sauce 4T, White Vinegar or Lemon Juice ¼ Cup. Blend all ingredients well.
- The squid paste is often already salty. Only need to add seasoning for the salinity of chicken.
- You could substitute with minced pork.

CC剛開始認識泰國時，在泰國的每天的早餐就是肉丸粥或海鮮粥(此2道料理收錄於《愛上CC老師的泰國菜》一書)，因為它們是CC每天最大的元氣來源。泰美味了！

When CC first learned to know Thailand, breakfast everyday is either meat ball congee or seafood congee (The two dishes were collected in the previous book, Falling in Love with CC's Thai Cuisine.), since they are the daily primary energy resources. Indeed, they are such "Thai" style delicacy!

泰式酸辣什錦河粉

Thai Style Spicy and Sour Assorted Wide Rice Noodle

材料：

泰國河粉適量、蝦 300g、透抽 300g、小里肌 400g、芹菜末 3 大匙、蔥花 3 大匙、香菜末 3 大匙、綠豆芽菜 100 大匙、檸檬 3 個、紅辣椒末、泰式綜合高湯 2000cc（參見 p.9）

醃料：

蒜末 1 大匙、檸檬葉（卡非萊姆）3 片（切碎）、紅蔥頭末 1 大匙、醬油 1 大匙、魚露 ⅔ 大匙、白砂糖 1 小匙，拌勻

調味料：

泰國辣椒膏 3 大匙、椰子糖（棕櫚糖）⅔ 大匙、魚露 3 大匙（視各人喜愛的鹹度）、檸檬汁（以各人喜愛之酸度）

做法：

1. 小里肌拌入醃料放置 2 小時後，放入鍋內煎熟取出切片。
2. 高湯注入湯鍋內放入調味料拌勻（檸檬汁除外），倒入海鮮煮熟，最後放入檸檬汁拌勻。
3. 河粉汆燙熟取出。
4. 將豆芽菜放於湯碗內，放上河粉擺上海鮮及做法(1)，撒上芹菜、香菜、蔥花、紅辣椒末，淋上湯汁即可。

Ingredient:

Thai Style Wide Rice Noodle some, Shrimp 300g, Squid 300g, Pork Tenderloin 400g, Finely-chopped Salary 3T, Finely-chopped Green Onion 3T, Finely-chopped Coriander 3T, Green Bean Sprouts 100g, Lemon 3, Finely-chopped Red Chili Pepper, Thai Style Mixed Soup Stock 2000cc (please see p.9)

Marinate:

Finely-chopped Garlic 1T, Lemon Leaf (Kaffir Lime) 3 pieces cut up, Finely-chopped Shallot 1T, Soy Bean Sauce 1T, Fish Sauce ⅔ T, Fine White Sugar 1t. Blend all ingredients evenly.

Seasoning:

Thai Chili Paste 3T, Coconut Sugar (Palm Sugar) ⅔ T, Fish Sauce 3T (subject to individual preference), Lemon Juice (subject to individual preference)

Procedure:

1. Blend pork tenderloin with marinate evenly and leave for 2 hours. Pan-fry till cooked and take the pan-fried pork out. Slice the pork tenderloin.
2. Add soup stock in a pot and blend with seasoning ingredients (except lemon juice) well. Add seafood to cook till done. Lastly add lemon juice to mix evenly.
3. Blanch wide rice noodle and take them out.
4. Put green bean sprouts on the bottom of a bowl. Add blanched wide rice noodle, seafood and Procedure (1). Sprinkle salary, coriander, finely-chopped green onion and finely-chopped red chili pepper. Pour soup to serve.

完美烹調寶典
Perfect Cooking Tips

- 可附上泰國辣椒粉及檸檬（切節狀）。
- 如無棕櫚糖（椰子糖）以白砂糖代替。

- You could also serve with Thai chili powder and lemon (cut in large pieces).
- If palm sugar (coconut sugar) is not available, substitute with fine white sugar.

喜歡泰國料理的人，一定不會漏掉這道，CC的一些朋友只要吃到這道一定會大叫幸福啊！

People who love Thai cuisine definitely won't miss this dish. Some of CC's friends always shout out their heartfelt satisfactions whenever they taste this rice noddle dish.

越南咖哩什錦河粉湯

Vietnamese Style Assorted Curry Rice Noodle Soup

這道料理學生們的回煮率最高。

This dish is among one of the most frequent cooking choices among students after class.

材料：

火鍋肉片（豬或牛）、蛤蠣 300g、透抽 1 枝、蝦 200g、越南河粉適量、洋蔥絲 ¼ 個、蔥花 1 枝、香菜末適量、紅蔥頭末 100g、檸檬 2 顆、越南高湯 2000cc（參見 p.10 頁）、九層塔

咖哩醬：

紅辣椒粉 ⅓ 大匙、紅辣椒末 1 大匙、蒜末 1 大匙、南薑末 1 大匙、蔥末 1 大匙、蝦米末 1 大匙、咖哩粉 4 大匙

調味料：

魚露 5 大匙、鹽適量、白砂糖 1 大匙

做法：

1. 鍋內放入 1½ 杯油待熱倒入紅蔥頭末炸至快呈金黃色，即馬上撈出瀝乾油（如已呈金黃色，再撈出會有苦味，因為溫度很高）為紅蔥頭酥；蝦米洗淨浸泡米酒 8 分鐘後取出擦乾水分。
2. 鍋入做法 (1) 的油 1 大匙待熱入蒜末爆香，至微黃再倒入蝦米爆香，入南薑末、蔥末、紅辣椒末炒約 2 分鐘，倒入咖哩粉（以小火炒）及紅辣椒粉，炒至咖哩粉香味溢出，再放入 3 大匙的紅蔥頭酥拌勻。
3. 越南高湯注入做法 (2) 拌勻，煮滾再放入蛤蠣及其他海鮮料煮至快熟，再放入調味料拌勻，最後放入肉片煮熟。
4. 河粉煮熟後取出瀝掉水分放入碗內，放入做法 (3) 撒上洋蔥絲、蔥花、香菜末，適量的紅蔥頭酥附上檸檬。

Ingredient:

Hotpot Meat Slices (either pork or beef), Clam 300g, Squid 1, Shrimp 200g, Vietnam Wide Rice Noodle some, Shredded Onion ¼, Finely-chopped Green Onion 1, Finely-chopped Coriander few, Finely-chopped Shallot 100g, Lemon 2, Vietnamese Style Soup Stock 2000cc (please see p.10), Basil Leaf

Curry Paste

Red Chili Pepper Powder ⅓ T, Finely-chopped Red Chili Pepper 1T, Finely-chopped Garlic 1T, Finely-chopped Galangal 1T, Finely-chopped Green Onion 1T, Finely-chopped small dried and shelled shrimps 1T, Curry Powder 4T

Seasoning:

Fish Sauce 5T, Salt few pinches, Fine White Sugar 1T

Procedure:

1. Add 1 ½ cups of oil to a pot to heat and add finely chopped shallot to deep fry till almost golden. Immediately take the deep-fried shallot out and drain off oil (If the shallot already turn golden brown in the pot, it would has bitter taste since the temperature of oil is very high.). Rinse the small dried shrimps and soak in rice wine for 8 minutes. Take them out and drain off water.
2. Add 1T of oil from Procedure (1) in a pot to heat and add finely chopped garlic to sauté till slight yellowish. Add small dried shrimps to sauté. Then add finely chopped galangal, green onion, and red chili pepper to stir for about 2 minutes. Add curry powder (low heat) and red chili pepper powder to stir till the smell of curry savor. Then add 3T of deep-fried shallot (Procedure (1)) to blend evenly.
3. Add Vietnamese style soup stock into Procedure (2) to mix well. Upon boiling add clam and other seafood ingredients to cook till almost done. Add seasoning to mix evenly. Lastly add meat slices to cook till well done.
4. Cook wide rice noodle well and take out to drain off water. Add in a bowl. Add Procedure (3) and sprinkle shredded onion, finely chopped green onion, coriander and some deep-fried shallot. Serve with lemon.

・咖哩不宜炒過久會苦澀味。

・Curry cannot be stirred too long otherwise it will has unpleasantly bitter taste.

泰國海鮮炒冬粉

Thai Style Stirred Green Bean Noodle with Seafood

材料：
寬冬粉3把、蝦120g、透抽150g、魚丸6個、洋蔥½顆（切絲）、綠韭菜3枝（切段）、芹菜3枝（切段）、紅辣椒2枝（切段）、蒜末1大匙、香菜末1大匙、蛋2顆（拌成蛋液）、蔥2枝、蝦米2大匙、米酒2大匙

調味料：
胡椒粉少許、魚露2大匙、醬油1大匙、白砂糖½大匙

做法：
1. 冬粉浸泡熱水約8分鐘後瀝乾水分備用。
2. 韭菜切段先分出白色部分以備爆香，芹菜及蔥切段，紅辣椒切段，蛋液打散，蝦米洗淨浸泡米酒6分鐘後瀝乾水分，魚丸對切，蝦去殼留尾巴去腸泥，透抽切條狀。
3. 鍋放入2大匙油待熱放入蒜末爆香，再入香菜末即入蝦米炒香，再入韭菜白及蔥白爆香，放入洋蔥絲略炒幾下，再入海鮮魚丸炒，並倒入蛋液炒再入冬粉拌炒，放入調味料拌勻最後倒進韭菜、蔥、芹菜段、紅辣椒拌炒勻即可盛盤。

Ingredient:
Wide Green Bean Noodle 3 Bunches, Shrimp 120g, Squid 150g, Fish Ball 6, Shredded Onion ½, Chinese Chive 3, Salary 3(segmented), Red Chili Pepper 2(segmented), Finely-chopped Garlic 1T, Finely-chopped Coriander 1T, Egg 2, Leek 2, Dried Shelled Small Shrimp 2T, Rice Wine 2T

Seasoning:
Pepper Powder few, Fish Sauce 2T, Soy Bean Sauce 1T, Fine White Sugar ½T

Procedure:
1. Soak green bean noodle in hot water for 8 minutes and drain off water.
2. Cut Chinese chive into segments. First take the white stalk part out for sauté use. Cut salary and leeks into segments. Cut red chili peppers into segments. Whisk eggs. Rinse the small dried shelled shrimps and soak them in rice wine for 6 minutes. Drain the soaked small shrimps. Cut fish balls into halves. Shell the shrimps with tails left and remove the intestine mud. Cut squid into strips.
3. Add 2T of oil in a pot to heat and add finely chopped garlic to sauté. Add finely chopped coriander and then the processed small dried shrimps to sauté. Then add the white stalk of Chinese chive and the white stem of green onion to sauté. Add shredded onion to slightly stir. Add seafood and fish balls to gently stir. Pour the whisked egg to stir before add green bean noodles. Blend them well. Add seasonings to mix evenly. Lastly add segmented Chinese chive, green onion, salary and red chili pepper to stir well. Place it to a plate

完美烹調寶典
Perfect Cooking Tips

- 韭菜頭白色部份是最佳的香料，例如在炒中式米粉時它是最佳的爆香食材。
- 炒時太乾可放入適量的泰式綜合高湯（參見p.9）或水。

- The white stalk of Chinese chive root could be ideal spice. For example, it is the best sauté ingredient for stirring Chinese rice vermicelli.
- If the moisture dries out too soon too much during stirring, you could add extra Thai Style Mixed soup stock (please see p.9) or water.

CC的家人都對冬粉情有獨鍾，尤其 CC 更是愛到不行，所以只要去到泰國一定會打聽一下那兒有好吃的冬粉，順便學習一下。

All CC' family members are particularly pond of green bean noodle and CC is the most enthusiasm one. Whenever visiting Thailand, CC always asks where to find the best noodle dishes and learn to do it.

越南香茅海鮮米粉湯

Vietnamese Style Lemon Grass Seafood Rice Vermicelli Soup

材料：

米粉適量、蝦 200g、蛤蠣 300g、透抽 1 隻、魚丸 150g、豆芽菜 100g、九層塔適量、檸檬 2 顆、蔥花 2 大匙、紅辣椒末適量

高湯：

雞骨 600g、豬骨 600g、香茅 5 枝（切段）、洋蔥 1 顆（切塊）、乾魷魚 150g、南薑 3 片、水 2500cc，熬煮 5 小時後過濾，使用壓力鍋 1 小時。

調味料：

魚露 5 大匙、鹽適量、白砂糖 1 大匙

做法：

1. 蝦去殼留尾巴由背部劃開取出腸泥；透抽切圈狀；蛤蠣用鹽水吐砂乾淨。
2. 高湯煮滾放入調味料拌勻，倒入海鮮及魚丸煮熟。
3. 米粉放入沸水中煮熟取出。
4. 取適量豆芽菜舖於器皿內放入適量的做法 (3)，倒入做法 (2) 撒上蔥花、紅辣椒及九層塔附上檸檬。

Ingredient:

Rice Vermicelli some, Shrimp 200g, Clam 300g, Squid 1, Fish Ball 150g, Green Bean Sprouts 100g, Basil Leaf few, Lemon 2, Finely-chopped Green Onion 2T, Finely-chopped Red Chili Pepper few

Soup Stock:

Chicken Bone 600g, Pork Bone 600g, Lemon Grass 5 (segmented), Onion 1(chucked), Dried Squid 150g, Galangal 3 slices, Water 2.5 liter,Braise the soup stock for 5 hours and then filter. If using a pressure cooker, cook for1 hour.

Seasoning:

Fish Sauce 5T, Salt few, Fine White Sugar 1T

Procedure:

1. Shell the shrimps with tails left. Cut the shrimps from back to remove the intestine mud. Cut squid into rings. Soak clams in salty water to expel sands.
2. Boil the soup stock and add seasoning to blend well. Add the processed seafood and fish balls to cook till well done.
3. Add rice vermicelli in another boiling pot to cook till done. Take the cooked rice vermicelli out.
4. Put some green bean sprouts on bottom of a container and then add proper proportion of Procedure (3). Then add Procedure (2). Sprinkle finely chopped green onion, red chili pepper and basil leafs. Serve with lemon.

完美烹調寶典 Perfect Cooking Tips

・此高湯可應用做火鍋鍋底，煮粥或做成越南海鮮湯。

・This soup stock could be used as hotpot base, congee or Vietnamese style seafood soup.

這道是 *CC* 在洛杉磯吃到的至今戀戀不忘，還好當時認真的提問一些問題，今天才能跟大家分享！

CC first ate this dish at LA and always keeps in mind constantly. Luckily CC did seriously ask many questions about this dish at that time so that today this dish could be shared in this book!

令人食指大動的
日韓料理

日式炸豬排套餐

Japanese Style Deep-fried Pork Chop Set Menu

炸豬排套餐是很多人想學的，尤其外面餐廳的售價不低，這一道套餐是 CC 最喜歡的味道跟大家分享。

The deep-fried pork chop set menu is one of the most enquired recipes among students in particular the price at restaurants is always very expensive. The taste of this recipe is CC's favorite and hereby shares with readers.

炸豬排

材料：

小里肌 600g、高麗菜切絲 120g、白飯適量、高筋麵粉 4 大匙、蛋 1 顆（拌成蛋液）、麵包粉 ¼ 杯

調味料：

黑胡椒粉適量、鹽適量（黑胡椒粉與鹽的比例為 2：1）、香蒜粉 1 小匙

味噌醬：

味噌 2 大匙、日式高湯 ¾ 杯（參見 P10）、昆布 1 片、小魚乾 ¼ 杯、砂糖 ½ 大匙，熬煮約 15 分鐘後過濾

沙拉醬：

白味噌 1½ 大匙、橄欖油 120cc、味醂 2 大匙、白砂糖 ⅓ 大匙打勻再加入檸檬汁 2½ 大匙、桔醬 1½ 大匙、柴魚醬油 1 大匙打勻

做法：

1. 小里肌肉切 1.5cm 厚度，以肉捶棒拍打使之鬆弛，撒上適量的黑胡椒粉、鹽和香蒜粉放置 10 分鐘。
2. 做法 (1) 均勻沾上麵粉→蛋液→麵包粉放入油溫 160℃ 油鍋炸至呈金黃色取出瀝乾油。

Ingredient:

Pork Tenderloin 600g, Shredded Cabbage 120g, Steam Rice some, Bread Flour 4T, Whisked Egg 1, Bread Powder ¼ Cup

Seasoning:

Black Pepper Powder and Salt few, Garlic Powder 1t

*The proportion between black pepper powder and salt is 2:1.

Miso Sauce:

Miso 2T, Japanese Style Soup Stock ¾ Cup (please see P.10), Kelp 1 piece, Dried Small Fish ⅕ Cup, Fine Sugar ½ T. Braise all ingredients for 15 minutes and then filter.

Salad Sauce:

White Miso 1 ½ T, Olive Oil 120cc, Mirin 2T, Fine White Sugar ⅓ T. Blend them well and then add lemon juice 2 ½ T, Kumquat Jam 1 ½ T, and Stockfish-flavored Soy Bean Sauce 1T. Mix well.

Procedure:

1. Cut pork tenderloin into 1.5cm thick and slap with tenderize to make it loose. Sprinkle few pinches of black pepper powder, salt and garlic powder and leave for 10 minutes.
2. Wrap Procedure (1) evenly with flour, whisked egg and then bread powder by the said order and deep fry in oil pot of 160° till golden brown. Take the deep-fried pork chops and drain off oil.

配菜

日式炒牛蒡

材料：

牛蒡 300g、白芝麻適量、白醋 2 大匙、水 600cc、香油 ½ 大匙

調味料：

醬油 1½ 大匙、柴魚醬油 1 大匙、味醂 1 大匙、砂糖 1 小匙

做法：

1. 牛蒡以刀背刨去外皮、切絲放入調理盆內，倒入水（水量需蓋過牛蒡）及白醋拌勻，浸泡 8 分鐘後洗淨，瀝乾水分。
2. 鍋內放入 1½ 大匙油待熱倒入牛蒡炒，並放入調味料及 3 大匙水炒至汁收乾，淋上香油拌勻，盛入盤內撒上白芝麻。

Side Dish

Japanese Style Stirred Burdock

Ingredient:

Burdock 300g, White Sesame some, White Vinegar 2T, Water 600cc, Sesame Oil ½ T

Seasoning:

Soy Bean Sauce 1 ½ T, Stockfish-flavored Soy Bean Sauce 1T, Mirin 1T, Fine Sugar 1t

Procedure:

1. Peel burdock with the back of knife. Shred the peeled burdock and put the shredded burdock in a basin. Blend well with water (water needs to cover over burdock) and white vinegar. Soak for 8 minutes. Then rinse and drain off water.
2. Add 1 ½ T of oil in a pot to heat and add the processed burdock to stir. Add seasoning and 3T of water to stir till water evaporated. Drip sesame oil and blend evenly. Place in a plate and sprinkle some white sesame.

日式蔬菜湯

材料：

高麗菜 200g（取剩下的）、肉片 100g（取小里肌剩下的小碎肉片）、日式高湯（參見 p.10）、蔥花適量、洋蔥絲 ⅓ 個、味噌 1 大匙、味醂 1 大匙

做法：

高麗菜剝片狀放入高湯內，並放入洋蔥、肉片，煮至洋蔥軟，放入味噌（味噌＋ 3 大匙水拌勻）及味醂拌勻以適量的鹽調味，盛入器皿內撒上蔥花。

Japanese Style Vegetable Soup

Ingredient:

Cabbage 200g (use what is left), Meat Slice 100g (use those small slices from tenderloin), Japanese Style Soup Stock (please see p.10), Finely-chopped Green Onion few, Shredded Onion ⅓, Miso 1T, Mirin 1T

Procedure:

Torn cabbage apart and put into soup stock. Add shredded onion and meat slices to cook till onion softened. Add miso (blend miso with 3T of water) and mirin to mix evenly. Add some pinches of salt as seasoning. Place in a container and sprinkle finely chopped green onion as final garnish.

完美烹調寶典 Perfect Cooking Tips

- 牛蒡浸泡醋水以防發黑，並去除草腥味。
- 可使用大里肌或雞胸肉，但均需注意火候以免過柴。

- Soak burdock in vinegar water to prevent from darken as well as to get rid of grassy smell.
- You could also use loin or chicken breast and be aware of heat control maturity in order not to overly cooked.

和風什錦飯

Japanese Style Seafood Casserole Rice

材料：

米 2 杯、鮭魚 600g、蛤蠣 600g、透抽 300g、蔥花 1 枝、昆布 1 片、蝦 200g、酒 1½ 大匙、日式高湯（參見 p.10）1 杯、蛤蠣汁 1 杯

調味料：

鹽適量

做法：

1. 蝦去殼留尾巴去腸泥；透抽切丁塊狀備用。
2. 鮭魚撒上少許鹽放置 15 分鐘後入鍋內，煎熟取出一半剝成小碎塊，另一半備用。
3. 蛤蠣放入鍋內倒入酒，蓋上鍋蓋煮至蛤蠣打開，取出後留下蛤蠣汁取出蛤蜊肉。
4. 米倒入鍋內注入高湯（高湯＋蛤蠣汁共 2 杯），鋪入少許鹽放上昆布，蓋上鍋蓋待冒煙改小火煮約 23 分鐘，放入做法 (1) 再煮約 7 分鐘（使用休閒鍋煮約 5 分鐘，再倒入做法 (1) 再煮 3 分鐘，移外鍋燜 6 分鐘）取出昆布。
5. 將做法 (2) 倒入做法 (4) 拌勻置於盤中，撒上蛤蠣肉，並將昆布切絲撒上，放上鮭魚片，最後撒上蔥花。

Ingredient:

Rice 2 Cups, Salmon Filet 600g, Clam 600g, Squid 300g, Finely-chopped Green Onion1, Kelp 1 piece, Shrimp 200g, Wine 1 ½ T, Japanese Style Soup Stock (please see p.10) 1 Cup, Oyster Juice 1 Cup

Seasoning:

Salt few

Procedure:

1. Shell the shrimps with tails left and remove the intestine mud. Cut squid into dices.
2. Sprinkle few pinches of salt onto salmon filet and leave for 15 minutes. Pan-fry till well done and take it out. Torn half of the pan fried salmon filet into small parts while keep the other half aside.
3. Add clams in a pot with wine. Cover with lid and cook till clams open. Take the cooked clams out and keep the clam juice. Shell the cooked clams and keep the clam meat. while keep the other half aside.
4. Add rice in a pot and pour soup stock(soup stock and calm juice together for 2 cups). Add few pinches of salt and kelp. Cover the lid to cook till smoking and turn to low heat to cook around 23 minutes. Add Procedure (1) to cook for another 7 minutes (If using HotPan, cook for 5 minutes and then add Procedure (1) to cook 3 minutes. Remove to outer cooker to simmer for 6 minutes.), then take the kelp out.
5. Pour Procedure (2) and pour into Procedure (4) to blend well. Place the rest half of Procedure (2) evenly onto Procedure (4). Top the clam meats onto it. Shred the kelp and put on the top. Put the rest half salmon filet. Sprinkle finely chopped green onion as final garnish.

・蛤蠣汁有鹹味調味時需注意。

・Clam juice is salty and be aware of this when seasoning.

這鍋飯讓很多人讚嘆。太有海味了！
This dish is widely proclaimed by many for the rich seafood flavor.

咖哩燉牛肉蓋飯

Curry Beef Stew Rice Bowl

材料：
牛腱 1kg、南瓜 300g（切塊狀）、牛番茄 2 顆（切節狀）、馬鈴薯 300g（切滾刀塊）、洋蔥 1 顆（切末）、薑末 ½ 大匙、蒜末 1½ 大匙、麵粉適量、咖哩粉 4 大匙、玉米粉適量、紅蘿蔔 1 條（切滾刀塊）、西式牛高湯（參見 p.9）1000cc、白飯適量

醃料：
咖哩粉 2 大匙、香蒜粉 1 小匙、鹽 ½ 大匙

做法：
1. 牛腱切 1.5cm 厚度，放入醃料拌勻，放置 30 分鐘後沾上麵粉，放入鍋中煎至呈金黃色取出。
2. 鍋內倒入 1½ 大匙油，待熱放入蒜末爆香，再放進洋蔥末及薑末，以小火炒至洋蔥呈現透明狀，即入咖哩粉炒至香味溢出。
3. 做法 (1) 倒入做法 (2) 拌炒勻，注入高湯燉煮約 60 分鐘，視牛腱快軟時放入紅蘿蔔及馬鈴薯煮約 5 分鐘，再入南瓜燉煮至南瓜熟，再入牛番茄，以適量的鹽調味，最後以玉米粉勾芡。
4. 白飯盛入器皿內淋上做法 (3)。

Ingredient:
Beef Shank 1kg, Pumpkin 300g (Chop in chucks), Beef Tomato 2 (Cut into large dices), Potato 300g(Chop in round chucks), Finely-chopped Onion 1, Finely-chopped Ginger ½ T, Finely-chopped Garlic 1 ½ T, Flour few, Curry Powder 4T, Corn Starch few, Carrot 1 (Cut in round chucks), Western Style Beef Soup Stock 1 liter(Please see p.9), Steam Rice some

Marinate:
Curry Powder 2T, Garlic Powder 1t, Salt ½ T

Procedure:
1. Cut beef shank into chucks in 1.5cm deep and blend evenly with marinate. Leave for 30 minutes. Wrap with flour and pan fry till golden brown. Take beef shank out.
2. Add 1 ½ T of oil to heat and add finely chopped garlic to sauté. Then add finely chopped onion and ginger to stir till onion turning transparent (low heat). Add curry powder to stir till savor.
3. Add Procedure (1) in Procedure (2) and blend and stir evenly. Pour soup stock to braise for 60 minutes. Add carrot and potato to cook for 5 minutes upon beef shank almost softened. Then add pumpkin to braise till well done. Lastly add beef tomato. Add few pinches of salt as seasoning and add corn starch to thicken.
4. Place steam rice in a container and pour Procedure (3)

完美烹調寶典 Perfect Cooking Tips

- 可改選用豬梅花肉、雞肉或牛肋條。
- 牛腱的軟度煮熟時間不一，需視牛腱的筋多少而定，以及牛隻來源為美國牛、澳洲牛或台灣牛。
- 咖哩粉不宜使用大火炒，也不宜炒過久，會有苦味。

- ou could substitute with pork shoulder meat, chicken or beef tenderloin.
- The cooking time of each beef shank is subject to change depending on how many tendons it has as well as the species, for example, American, Australian or Taiwanese beef.
- It is suggested not to stir curry powder wit high heat or in long period of time otherwise it will create bitter taste.

隔夜更好吃的咖哩燉牛肉。

Leave this curry beef stew overnight will create more flavor and taste.

京都燒肉佐山藥蓋飯

Kyoto Braised Pork Belly with Chinese Yam Rice Bowl

材料：

五花肉 1kg、山藥 200g、蔥花 2 枝、白飯適量、蔥 5 枝、薑 3 片、蒜頭 8 粒、八角 2 粒、紅辣椒 2 枝、醬油 6 ～ 8 大匙、冰糖 1 大匙、濃口醬油 2 大匙、水 1½ 杯

煮料：

味噌 1½ 大匙、味醂 1½ 大匙、砂糖 1 大匙、酒 1 大匙、湯汁（滷肉的汁）

做法：

1. 五花肉切約 2.5cm 的厚度，長度約 10cm。
2. 鍋待熱放入做法 (1) 皮先朝鍋底（不放油）煎至呈金黃色，再翻面煎，煎至呈金黃色取出，放入蔥、薑、蒜頭爆香再放入煎至金黃色的肉並放入八角、冰糖、紅辣椒、濃口醬油、水，燉煮約 40 分鐘（如使用壓力鍋，醬油 2 大匙，濃口醬油 1 大匙（不放水），待上升二條紅線改小火煮 12 分鐘），取出肉塊湯汁過濾。
3. 將煮料及做法 (2) 的湯汁拌勻再倒入做法 (2) 的肉燉煮約 20 分鐘（使用壓力鍋，待上升二條線改小火煮 10 分鐘）。
4. 白飯盛於器皿內放上肉塊淋上少許湯汁，再淋上磨好的山藥泥，撒上蔥花即可。

Ingredient:

Pork Belly 1kg(1000g), Chinese Yam 200g, Finely-chopped Green Onion 2, Steam Rice some, Leek 5, Ginger Slices 3, Garlic 8, Aniseed 2, Red Chili Pepper 2, Soy Bean Sauce 6 to 8T, Crystal Sugar 1T, Dark Shoyu (Thick Japanese Soy Bean Sauce)2T, Water 1½Cups

Cooking Sauce

Miso 1 ½ T, Mirin 1 ½ T, Fine White Sugar 1T, Wine 1T, Soup(braised pork belly juice)

Procedure:

1. Cut pork belly into 10cm long and 2.5cm thick.
2. Heat a pot and add Procedure (1) with skin side facing down (without oil) to pan-fry till golden. Then turn the other side facing down to pan-fry till golden as well. Take the pan-fried pork belly out. Add 5 leeks, 3 slices of ginger, 8 garlics to stir till savor, then add the golden pan-fry pork, aniseed, crystal sugar, red chili pepper, dark shoyu and water to braise for about 40 minutes (if using a pressure cooker, add 2T of soy bean sauce and 1T of dark shoyu (without water) instead. Turn to low heat to cook 12 minutes when it rises to two red strips.) Take the braised pork belly out and filter the juice.
3. Blend cooking sauce and juice made from Procedure (2) well first and then add braised meat from Procedure (2) to braise for another 20 minutes (If using a pressure cooker, turn to low heat to cook 10 minutes when it rises to two red strips.).
4. Place steam rice in a container. Add some braised meat and drip few meat juice. Then add smashed Chinese Yam paste on top. Sprinkle with finely chopped green onion as final garnish.

完美烹調寶典
Perfect Cooking Tips

- 白豬肉烹調時間較短，本書所使用的為黑豬肉。
- 日本濃口醬油較鹹，可使用老抽代替，老抽指顏色深的醬油，南北雜貨店或台北 SOGO 百貨（復興店）均有販售。

- The white-skin pork requires less cooking time and yet in this recipe black skin pork is used instead.
- The dark shoyu is generally very salty and it is suggested to replace with dark soy bean sauce. Dark soy bean sauce refers to dark-colored soy bean sauce and is available at dried food and grocery stores or Taipei Sogo Department Store (Fuxing Branch).

山藥泥配上熱騰騰的白飯，淋上醬汁或醬油，撒上蔥花是令人著迷的吃法。再搭配上燒肉（滷肉）那種滋味實在太超過了！太想多吃幾碗，這是CC學生對這道蓋飯的評價！

Cool smashed Chinese yam paste with smoking-hot steam rice, dripping sauce or soy bean sauce with some finely-chopped green onion is such fascinating way to enjoy a rice bowl. Furthermore, topping with finely-braised pork is way too far beyond speechless! "Always want to have another bowl or more than one bowl of this," is responses from CC's students on this dish!

大阪味噌豬排蓋飯

Osaka Miso Sauce Deep-fried Pork Chop Rice Bowl

材料：

大里肌或小里肌 600g、白飯適量、洋蔥 ½ 個切絲、蔥花 2 大匙、柳松菇 120g、蔥花適量、日式高湯 ¼ 杯（參見 p.10）

醃料：

味噌（白）1½ 大匙、味醂 ⅔ 大匙、薑泥 ½ 大匙、蔥末 1 大匙、蒜末 1½ 大匙、蘋果泥 1 大匙、洋蔥泥 1 大匙、香油 ½ 大匙、黑胡椒粉 1 小匙、醬油 2 大匙、柴魚醬油 1 大匙、米酒 ½ 大匙，拌勻

調味料：

柴魚醬油 1 大匙、醬油 1 大匙、味醂 ⅔ 大匙

做法：

1. 里肌肉切約 1cm 的厚度，用肉搥棒拍打使之鬆弛，放入醃料拌勻，醃漬 2 小時。
2. 鍋內倒入 1 大匙油待熱放入做法 (1) 煎熟取出。
3. 洋蔥絲倒入做法 (2) 的鍋內炒幾下，倒進柳松菇炒，放入高湯及調味料拌炒勻至熟。
4. 白飯放入器皿內舖上做法 (3) 再放上做法 (2)，放上蔥花。

Ingredient:

Pork Tenderloin or Pork Belly 600g, Steam Rice some, Shredded Onion ½, Finely-chopped Green Onion 2T, Brown Swordbelt 120g, Finely-chopped Green Onion few, White Sesame few, Japanese Style Soup Stock ¼ Cup (please see p.10)

Marinate:

Miso (white) 1 ½ T, Mirin ⅔ T, Ginger Paste ½ T, Finely-chopped Green Onion 1T, Finely-chopped Garlic 1 ½ T, Apple Paste 1T, Onion Paste 1T, Sesame ½ T, Black Pepper Powder 1t, Soy Bean Sauce 2T, Stockfish-flavored Soy Bean Sauce 1T, Rice Wine ½ T. Blend all ingredients well.

Seasoning:

Stockfish-flavored Soy Bean Sauce 1T, Soy Bean Sauce 1T, Mirin ⅔ T

Procedure:

1. Cut pork tenderloin or pork belly into 1cm thick and slap with tenderizer to get soft texture. Blend with marinate well and leave for 2 hours.
2. Add 1T of oil in a pot to heat and add Procedure (1) to pan-fry till cooked. Take the pan-fried meat out.
3. Add shredded onion into Procedure (2) to slightly stir. Add brown swordbelt to stir. Then add soup stock and seasoning to blend and stir evenly. Cook till well done.
4. Add steam rice in a container and add Procedure (3) on top of rice. Then add Procedure (2) on top of Procedure (3). Sprinkle white sesame, shredded green onion as final garnish.

・可選用去骨雞腿肉代替。

・You could substitute with boned drumsticks.

喜歡比較道地口味的醃料，可運用在烤肉上哦！
For those who like? marinate, this could apply to BBQ as well!

照燒豬排蓋炒飯

Teriyaki Deep-fried Pork Chop Fried Rice Bowl

材料：

里肌肉排 4 片約 600g、白飯 4 碗、白芝麻少許、奶油 ½ 大匙、芹菜末 4 大匙、紅蘿蔔末 4 大匙、洋蔥末 ¼ 杯

調味料：

米酒 2 大匙、白胡椒粉少許、醬油適量、鹽少許

照燒醬：

醬油 5 大匙、老抽 2 大匙、味醂 2 大匙、白砂糖 3 大匙、米酒 2 大匙、柴魚精 1 大匙，放入鍋內以小火煮約 10 分鐘（需常攪拌），放涼後會型成稠狀，存放於冰箱冷藏。

做法：

1. 里肌肉片以肉搥棒拍打使之鬆弛。
2. 刷子沾上照燒醬刷上做法 (1) 正反面放置 10 分鐘。
3. 鍋內放入 1 大匙油待熱放入做法 (2) 煎，邊煎邊刷約 3 次（兩面刷各 3 次）至熟，最後放入 ½ 大匙奶油使肉片更香更油亮。
4. 鍋內放入 2½ 大匙油待熱倒入洋蔥末炒香，再入紅蘿蔔末炒並放入一半的芹菜末炒香，倒入白飯由鍋邊淋上米酒，放入其他調味料炒勻，最後再撒上剩餘的芹菜末拌炒勻盛入盤內。
5. 將煎好的豬排舖在做法 (4) 上撒上白芝麻。

Ingredient:

Pork Tenderloin 4 pieces about 600g, Steam Rice 4 Bowls, White Sesame few, Butter ½ T, Finely-chopped Salary 4T, Finely-chopped Carrot 4T, Finely-chopped Onion ¼ Cup

Seasoning:

Rice Wine 2T, White Pepper Powder few, Soy Bean Sauce some, Salt few

Teriyaki Sauce

Soy Bean Sauce 5T, Dark Soy Bean Sauce 2T, Mirin 2T, Fine White Sugar 3T, Rice Wine 2T, Stockfish Essence 1T. Put all ingredients in a pot to cook for 10 minutes with low heat while continuing stirring. Cool down to get thickened and keep cold in the refrigerator.

Procedure:

1. Slap pork tenderloin with tenderizer to make it loose.
2. Brush teriyaki sauce onto the two sides of Procedure (1) and leave for 10 minutes.
3. Add 1T of oil in a pot to heat and pan-fry Procedure (2) while brushing more teriyaki sauce on both sides of meat (each side should be brushed three times) till cooked. Then add ½ T of butter to glaze meats and bring over strong savor.
4. Add 2 ½ T of oil in a pot to heat and add finely-chopped onion to sauté. Add finely-chopped carrot to stir and then add half of finely-chopped salary to sauté. Add steam rice and drip rice wine along the pot edge. Then add other seasoning to stir and blend evenly. Lastly add the rest half of finely-chopped salary to blend well. Place in a plate.
5. Place the pan-fried pork chops on top of Procedure (4) and sprinkle white sesame as final garnish.

完美烹調寶典
Perfect Cooking Tips

- 可選用去骨雞腿或去骨去皮的魚肉或牛排。
- 照燒醬可用在大阪燒煎餅。

- You could also use boned drumsticks, boned and skinned fish filet, or beef steak.
- This teriyaki sauce could be used for okonomiyaki (Osaka style pancake).

CC在唸書的時代最喜歡到一家有名的西餐廳，在當時是很潮的，因為喜歡吃這道飯料理，所以存錢為了想吃它，後來想盡辦法問到如何烹調。

During CC's school years, there was a famous Western style restaurant which is quite fashionable and CC always loved to go. Since fond of this dish, CC saved money just in order to enjoy it. Later on CC made great efforts to ask for the recipe and cooking method.

北海道櫻花蝦蛤蠣飯

Hokkaido Sergestid Shrimp and Clam Rice Bowl

材料：

櫻花蝦 ½ 杯、綠竹筍 2 支或沙拉筍 1 包、蛤蠣 1 斤、昆布 1 片、米 2 杯、日式高湯（參見 p.10）1 ¾、蔥花 2 大匙、米酒 2 ½ 大匙

調味料：

鹽適量

做法：

1. 櫻花蝦洗淨浸泡米酒 1 ½ 大匙，倒入水蓋住櫻花蝦浸泡約 6 分鐘後瀝乾水分，蛤蠣以鹽水吐砂。
2. 綠竹筍連同外殼注入洗米水煮至筍熟即可，取出再剝去外殼切丁。
3. 蛤蠣倒入鍋內放入 1 大匙酒蓋鍋蓋，煮至蛤蠣打開取出湯汁與高湯一起總共 2 杯湯汁。
4. 鍋內放入 ½ 大匙油待熱入蔥花爆香，倒入櫻花蝦炒香即入米（小火），炒約 1 分鐘倒入筍丁炒，注入 2 杯高湯（做法 (3) 的湯汁）放入適量的鹽拌勻，放上 1 片昆布蓋上鍋蓋，待冒煙改小火煮 25 ～ 30 分鐘（使用休閒鍋煮，待冒煙改小火煮 6 分鐘，移外鍋燜 6 分鐘），取出昆布。
5. 盛入器皿內放上蛤蠣肉，中間放上蔥花。

Ingredient:

Sergestid Shrimp ½ Cup, Green Bamboo Shoots 2 or Processed Bamboo Shoots for Salad Use 1 bag, Clam 600g, Kelp 1 piece, Rice 2 Cups, Japanese Style Soup Stock 1 ¾ Cup (please see p.10), Finely-chopped Green Onion 2T, Rice Wine 2 ½ T

Seasoning:

Salt few pinches

Procedure:

1. Rinse Sergestid shrimps and soak with 1 ½ cups of rice wine and water till fully covered for about 6 minutes. Drain off water. Clams put into salty water to expel sands.
2. Cook green bamboo shoots with outer skin together in rice-rinsing water till well done. Then take them out to peel and cut into dices.
3. Add clams in a pot with 1T of wine then covered. Cook till clams open and take the juice out to mix with soup stock as 2 cups.
4. Add ½ T of oil to heat and add finely chopped green onion to sauté. Add Sergestid shrimps to sauté and then add rice upon savor (low heat). Stir about 1 minute later add bamboo shoot dices to stir slightly. Pour 2 cup of soup stock(soup from Procedure 3) with few pinches of salt to blend evenly. Place a piece of kelp on top and cover the pot. Turn to low heat upon smoking and cook for another 25 to 30 minutes (If using HotPan, turn to low heat to cook for 6 minutes upon smoking. Then remove to outer cook to simmer for another 6 minutes.). Remove the kelp.
5. Place into a container. Put clam meats on top. Sprinkle finely chopped green onion in the center.

完美烹調寶典 Perfect Cooking Tips

- 蛤蠣的湯汁已有鹹味，調味時需留意。
- 昆布不洗僅以紙巾擦拭（保有其甜味）。

- The clam juice is salty, so be aware when adding seasoning.
- Do not rinse the kelp. Wipe with paper towel (to preserve the natural sweet flavor).

CC 常在日本料理店，或中式料理店吃到櫻花蝦飯，但總是不太滿足，所以 *CC* 多加了蛤蠣和竹筍，結果大受歡迎。

CC often eats Sergestid shrimp rice bowl at either Japanese cuisine restaurants or Chinese diners and is not very contented with the taste. Therefore CC adds clam and bamboo shoots. It turns out to be extremely popular.

日式牛五花蓋飯

Japanese Style Beef Briskets Rice Bowl

材料：

牛五花或牛肉片 300g、柳松菇或美白菇 150g、洋蔥絲 ½ 個、白飯適量、奶油 1 大匙、日式高湯 ⅓ 杯（參見 p.10）、蔥絲少許

醃料：

洋蔥泥 1 大匙、香蒜粉 1 小匙、醬油 1 大匙

調味料：

味醂 1 大匙、柴魚醬油 2 大匙、醬油 1½ 大匙

做法：

1. 牛肉拌入醃料拌勻置 15 分鐘。
2. 奶油倒入鍋內待溶化，倒入洋蔥炒至洋蔥呈現軟化，放入柳松菇略炒幾下倒入做法 (1) 拌炒，並放入高湯及調味料快速拌煮即可。
3. 白飯盛入器皿鋪上做法 (2) 撒蔥絲。

Ingredient:

Beef Briskets or Beef Slices 300g, Brown Swordbelt or Hongshi Mushroom 150g, Shredded Onion ½, Steam Rice some, Butter 1T, Japanese Style Soup Stock ⅓ Cup (Please see p.10), Shredded Green Onion few

Marinate:

Onion Paste 1T, Garlic Powder 1t, Soy Bean Sauce 1T

Seasoning:

Mirin 1T, Stockfish flavor Soy Bean Sauce 2T, Soy Bean Sauce 1 ½ T

Procedure:

1. Blend beef with marinate and leave for 15 minutes.
2. Melt butter in a pan and add shredded onion to stir till softened. Add brown swordbelt to stir slightly and then add Procedure (1) to blend and stir. Add soup stock and seasoning to stir and cook quickly.
3. Put steam rice in a container and top with Procedure (2). Sprinkle shredded green onion.

・可使用豬五花肉片或火鍋肉片。

・You could substitute with pork, pork belly or hotpot meat slices.

全家都喜歡又簡單的一道
日式牛丼飯。
A simple Japanese style beef rice bowl fond by the whole family.

九州田園燒肉蓋飯

Kyushu Countryside Braised Pork Rice Bowl

材料：

五花肉 1kg、地瓜 1kg、蔥末適量、白飯適量、水 2 杯、紅蘿蔔 300g

滷料：

柴魚精 1 大匙、味醂 ¼ 杯、砂糖 3 大匙、醬油 ½ 杯、濃口醬油 ⅓ 杯

做法：

1. 五花肉切 2cm 厚度，長約 8cm；地瓜、紅蘿蔔削皮切滾刀塊狀。
2. 鍋待熱，五花肉皮朝鍋底煎至呈金黃色，熄火倒入滷料及水和紅蘿蔔拌勻再開火燒滷約 1 小時（使用壓力鍋不放水，醬油和濃口醬油各 2 大匙共 4 大匙，味醂 1 大匙，糖 2 大匙，待上升二條紅線改小火 18 分鐘）。
3. 打開鍋蓋再倒入地瓜煮至地瓜熟。
4. 白飯盛入器皿內放上肉及地瓜和紅蘿蔔，淋上滷汁撒上蔥花即可。

Ingredient:

Pork Belly 1kg, Sweet Potato 1kg, Finely-chopped Green Onion and Steam Rice some, Water 2 Cups, Carrot 300g

Braising Sauce

Stockfish Essence 1T, Mirin ¼ Cup, Sugar 3T, Soy Bean Sauce ½ Cup, Dark Shoyu (Thick Japanese Soy Bean Sauce) ⅓ Cup

Procedure:

1. Cut pork belly into strips of 2 cm deep and 8cm long. Peel sweet potato and carrot and chop into round chucks.
2. Heat a pan and put the skin side of pork belly facing the pan to pan fry till golden brown. Turn off the heat to add braising sauce, water and carrot. Blend evenly and then turn on the heat again to braise for 1 hour (If using a pressure cooker, no need to add water. Soy bean sauce and dark shoyu are totally 4T, mirin 1T, Sugar 2T. Turn to low heat to cook for 18 minutes when it rises to two red strips.).
3. Open the lid to add sweet potato to cook till well done.
4. Place steam rice into a bowl and top with braised pork, sweet potato and carrot. Drip braising sauce. Sprinkle finely chopped green onion.

完美烹調寶典
Perfect Cooking Tips

- 白毛豬烹調時間會較快。
- 濃口醬油指顏色深的醬油，日本進口的會偏過鹹，可用老抽代替（顏深的醬油）。
- 醬油品牌的鹹度不一需注意。

- White skin pork tends to be cook faster.
- Dark Shoyu refers to dark color soy bean sauce. Some Japanese imported brands might be too salty and you could substitute with dark soy bean sauce (another type of dark color soy bean sauce).
- The salinity of different brands of soy bean sauce is subject to change. Be aware when using.

很多人告訴 *CC*，從來不知道地瓜和豬肉燒滷出來的味道是這麼的合，這麼無法想像的美味。

Many people told CC that they never know that sweet potato braising together with pork could be so delicious. The taste is unimaginable and yet very delicate.

東京豆奶海鮮炊飯

Tokyo Soy Milk Seafood Japanese Style Casserole Rice Bowl

材料：
蝦200g、透抽1隻、蛤蠣300g、牡蠣200g、鮮干貝150g、昆布1片、無糖豆漿1杯、日式高湯1杯（參見p.10）、米2杯、蔥花1枝、麵粉2大匙、酒2大匙

做法：
1. 蝦去殼留尾巴，透抽切圈狀，牡蠣用麵粉及1大匙酒輕拌勻再洗淨，瀝乾水分。
2. 蛤蠣放入鍋內倒入1大匙酒，煮至蛤蠣打開取出肉，留取湯汁。
3. 米倒入鍋內（小火）炒至呈象牙白色，注入高湯（連蛤蠣汁及高湯共1杯）和豆漿1杯，撒入適量的鹽拌勻，放上一片昆布蓋上鍋蓋，小火煮至飯約7分熟，倒入海鮮再燜煮至海鮮熟（使用休閒鍋或雙享鍋煮約5分鐘後，放入海鮮再煮1分鐘移入外鍋燜6分鐘即可）。
4. 將昆布取出盛入碗內再放入蔥花。

Ingredient:
Shrimp 200g, Squid 1, Clam 300g, Oyster 200g, Raw Scallop 150g, Kelp 1 piece, Sugar-free Soy Milk 1 Cup, Japanese Style Soup Stock 1 Cup (please see p.10), Rice 2 Cups, Finely-chopped Green Onion 1, Flour 2T, Wine 2T

Procedure:
1. Shell the shrimps with tails left and remove the intestine mud. Cut squid into rings. Rinse oyster gently with flour and 1T of wine and then drain off water.
2. Add clams in a pot with 1T of wine. Cook till clams open. Take the cooked clams out and keep the clam juice. Shell the cooked clams and keep the clam meat.
3. Add rice in a pot to stir (low heat) till turning ivory white. Add soup stock (mix clam juice and soup stock in 1 cup), soy milk 1 cup and few pinches of salt to blend evenly. Add a piece of kelp and cover with the lid to cook with medium to low heat till medium raw. Then add seafood to cook with lid covered till seafood cooked (If using HotPan, cook 5 minutes and add seafood to cook 1 more minute. Remove to outer cooker to simmer for another 6 minutes.).
4. Take the kelp out, put in a bowl and sprinkle with finely chopped green onion.

- CC常將昆布取出後切絲再放回鍋內，不浪費食材。
- 可使用自己喜愛的海鮮。

- CC often shreds the cooked kelp and put back to the pot in order not to waste food ingredient.
- You could use any kinds of seafood you prefer.

以豆漿和海鮮做成的炊飯是很健康又有清新味道的飯類料理，深受主婦們喜愛。

The Japanese style casserole rice bowl made of soy milk and assorted seafood is very nutritious and of fresh taste. This rice dish is profoundly fond by housewives.

築地海鮮粥

Tzukiji Seafood Congee

材料：

米 2 杯、日式高湯 2300cc（參見 p.10）、蛤蠣 600g、透抽 500g、蝦 300g、鮮干貝 150g、杏鮑菇 200g、蔥花適量、柴魚醬油 ¼ 杯、鹽適量

做法：

1. 蝦去殼去腸泥；透抽切圈狀或丁狀；杏鮑菇切丁狀；干貝一顆切 3 塊；蛤蜊放入鹽水吐砂。
2. 米倒入鍋內（不放油）炒至米鬆鬆的約 1～1 分半鐘（小火），倒入柴魚醬油拌勻，注入高湯煮至呈現粥狀，冒煙改小火約 20～25 分鐘。（壓力鍋待上升二條紅線改小火 6 分鐘）
3. 杏鮑菇及蛤蠣倒入做法 (2) 內煮約 1 分鐘，放入透抽及蝦，再入干貝煮至海鮮熟，以適量的鹽調味拌勻，盛入器皿內撒上蔥花。

Ingredient:

Rice 2 Cups, Japanese Style Soup Stock 2300cc (please see p.10), Clam 600g, Squid 500g, Shrimp 300g, Fresh Scallop 150g, King Oyster Mushroom 200g, Finely-chopped Green Onion few, Stockfish-flavored Soy Bean Sauce ¼ Cup, Salt few pinches.

Procedure:

1. Shell the shrimps and remove the intestine mud. Cut squids into rings or dices. Cut king oyster mushroom in dices. Cut one scallop into 3 pieces. Put clams into salty water to expel sands.
2. Add rice into pot (without oil) to stir one and half minutes till loosen (low heat). Add stockfish-flavored soy bean sauce to blend well. Add soup stock to braise till congee status. Turn to low heat upon smoking and cook for 20 to 25 minutes. (If using a pressure cooker, turn to low heat to cook 6 minutes when it rises to two red strips.)
3. Add king oyster mushroom and clams into Procedure (2) and cook for 1 minute. Then add squids and shrimps, and lastly scallops to cook till all seafood done. Blend evenly with few pinches of salt as seasoning. Place in a container and sprinkle finely chopped green onion as final garnish.

完美烹調寶典 Perfect Cooking Tips

- 可放入自己喜愛的海鮮。
- 蛤蠣有鹹度，調味時須多加斟酌醬汁用量。

- You could use any kinds of seafood as you prefer.
- Clam is already salty. Be aware of this when making seasoning.

粥的做法很多種，這是 CC 最喜歡的味道之一。這本書介紹的粥都是 CC 愛的粥品，也應說最受所有同學 (學生) 的喜愛。

There many ways to make congee and this one is CC's favorite. It should be one of the most favorite ones since all congees in this book are very much fond by CC. They are also the most popular among CC's students.

日式咖哩烏龍麵

Japanese Style Curry Udon

材料：

烏龍麵 3 包、蝦 9 隻、透抽 1 隻、蛤蠣 200g、蛋 3 顆、蔥花適量、日式高湯 1200cc（參見 p.10）、咖哩醬 6 大匙（參見 p.8）、月桂葉 1 片、番茄粒 ⅓ 罐

調味料：

柴魚醬油 ¼ 杯、鹽適量、味醂 1½ 大匙

做法：

1. 咖哩醬倒入鍋內，高湯慢慢倒入以打蛋器拌勻，放入搗碎的番茄粒罐頭及月桂葉約煮 10 分鐘。
2. 調味料放入做法 (1) 內拌勻，即入蛤蠣、烏龍麵、蝦和透抽煮至快熟，放入蛋撒上蔥花。

Ingredient:

Udon 3 bags, Shrimp 9, Squid 1, Clam 200g, Egg 3, Finely-chopped Green Onion few, Japanese Style Soup Stock 1200cc (please see p.10), Curry Paste 6T (please see p.8), Bally Leaf 1, Tomato Dices ⅓ can

Seasoning:

Stockfish-flavored Soy Bean Sauce ¼ Cup, Salt few, Mirin 1 ½ T

Procedure:

1. Add curry paste in a pot and slowly pour soup stock while blending evenly with egg-whisker. Add canned tomato dices and bally leaf to cook about 10 minutes.
2. Add seasoning in Procedure (1) and blend well. Add clam, udon, shrimps and squids to cook till almost done. Add egg and sprinkle finely chopped green onion.

・使用冷凍烏龍麵口感會較 Q。

・The frozen udon has more chewy texture.

很多人喜歡咖哩，但要吃到咖哩烏龍麵卻不常有，所以這道咖哩烏龍麵是 *CC* 的學生們強力要求 *CC* 一定要寫在這本書上，因為他們覺得太美味了。

Curry is widely fond by many people and yet it is not that often to enjoy curry udon. Thus this curry udon is upon strongly requests from CC's students to be showed in this book because they really think this dish is too delicious not to share.

韓式軟殼蟹蟹肉炒飯

Korean Style Soft-Shell Crab and Crab Meat Fried Rice

材料：

軟殼蟹 4 隻、蒜末 1 大匙、白飯 4 碗、泡菜 150g、蟹肉或蟹管肉 100g、酥脆粉 4 大匙、蛋 4 顆、培根 3 片（切絲）、洋蔥末 3 大匙、泡菜（切丁）、香菜末 1 大匙、蔥花 2 大匙

醃料：

酒 1 大匙、蔥 1 枝（拍扁）、薑 2 片（拍扁）、白胡椒粉少許

調味料：

米酒 1 大匙、醬油 2 大匙、鹽適量、香油 1 大匙、白胡椒粉少許

做法：

1. 軟殼蟹去鰓瀝乾水分。
2. 醃料拌勻放入做法 (1) 再拌勻靜置 20 分鐘。
3. 酥脆粉加入適量的水拌勻呈稠狀放入做法 (2)，平均沾勻入油溫 160°C 油鍋炸至呈金黃色取出瀝掉油。
4. 鍋內放入 1 大匙油，放入培根炒至酥脆，倒入蒜末、洋蔥末爆香，倒入蛋液拌炒勻即入白飯，由鍋邊淋上米酒拌勻再放入調味料拌勻（香油除外）。倒入泡菜拌炒均勻，最後撒上香菜末及蔥花淋上香油拌均勻，即可盛入盤中再放上炸好的軟殼蟹。

Ingredient:

Soft Shell Crab 4, Finely-chopped Garlic 1T, Steam Rice 4 bowls, Kinchi (Korean pickled vegetable) 150g, Crab Meat or Compressed Crab Meat 100g, Crispy Powder 4T, Egg 4, Bacon 3 slice (shredded), Finely-chopped Onion 3T, Diced Kinchi, Finely-chopped Coriander 1T, Finely-chopped Green Onion 2T

Marinate:

Wine 1T, Leek 1 (crush flat), Ginger 2 slices (crush flat), White Pepper Powder few

Seasoning:

Rice Wine 1T, Soy Bean Sauce 2T, Salt, Sesame Oil 1T, White Pepper Powder few

Procedure:

1. Remove the gill of soft shell crabs and drain off water
2. Mix the marinate ingredients well and then blend with Procedure (1). Leave for 20 minutes.
3. Add water of proper proportion into crispy powder and mix till thickened. Put Procedure (2) in it to wrap evenly. Deep fry in an oil pot at 160°C till golden crispy. Take soft shell crabs out and drain off oil.
4. Add 1T of oil in a pot to pan fry bacon till crispy. Then add finely chopped garlic and onion to sauté. Add whisked eggs to blend well and pour steam rice in. Drip rice wine around the edge of the pot. Add seasoning to blend well (except sesame oil). Finally add kinchi to stir and blend. Sprinkle finely chopped coriander and green onion. Drip sesame oil to mix evenly. Place into a plate and put deep fried soft-shell crabs on top.

完美烹調寶典
Perfect Cooking Tips

・軟殼蟹本身已有鹹味不需放鹽。
・酥脆粉超市有售，或使用地瓜粉或太白粉。

・Soft shell crab is already salty and thus no need to add salt.
・Crispy powder is available at supermarkets or you could substitute with sweet potato starch or potato starch.

韓國風味的蓋飯，以軟殼蟹作為主軸，呈現出華麗美味的風格。

This Korean-flavor bowl dish is accented with soft shell crab and thus presents a luxurious delicious style.

韓國海鮮鍋飯

Korean Seafood Casserole Rice

材料：

蛤蠣300g、牡蠣200g、蝦200g、透抽1隻、韓式高湯1杯（參見p.10）、米2杯、泡菜150～200g（切絲）、培根3片（切絲）、薑末1大匙、蒜末1大匙、奶油1大匙、香油2大匙、白酒1½大匙、洋蔥末4大匙、蔥花2大匙、香菜適量、醬油1大匙、白芝麻少許

做法：

1. 透抽切圈狀；蝦去殼留尾巴去腸泥。
2. 鍋入奶油待融化，放洋蔥末爆香，入蛤蠣炒幾下，入透抽、蝦、牡蠣，淋上白酒使之蒸發，淋上泡菜汁拌勻煮至海鮮熟，取出湯汁備用。
3. 鍋倒入香油1½大匙，放進蒜末、薑末、洋蔥末爆香，即入培根絲炒至呈現焦黃色，倒入米以小火炒約1分鐘，淋上1大匙醬油拌勻，注入做法(2)的湯汁及高湯（湯汁及高湯共2杯），拌勻以適量的鹽調味，蓋上鍋蓋以中小火煮，待冒煙改小火煮25～30分鐘（如使用較好材質的休閒鍋約煮6分鐘，熄火移入外鍋燜6分鐘）。
4. 做法(2)的海鮮取一半倒入做法(3)並倒入泡菜絲拌勻，再舖上剩下一半的海鮮，淋上香油約½大匙，撒上蔥花、香菜末、及白芝麻。

Ingredient:

Clam 300g, Oyster 200g, Shrimp 200g, Squid 1, Korean Style Soup Stock 1 Cup (please see p.10), Rice 2 Cups, Kinchi (Korean Style Pickled Vegetables) 150 to 200g(shredded), Bacon 3 slices (shredded), Finely-chopped Ginger 1T, Finely-chopped Garlic 1T, Butter 1T, Sesame Oil 2T, White Wine 1 ½ T, Finely-chopped Onion 4T, Finely-chopped Green Onion 2T, Coriander few, Say Bean Sauce 1T, White Sesame few

Procedure:

1. Cut squid into rings. Shell the shrimps with tails left and remove the intestine mud.
2. Melt butter in a pot and add finely chopped onion to sauté. Then add clams to slight stir. Add squid, shrimps, and oysters. Drip white wine to cook till evaporated. Drip kinchi juice to blend evenly and cook till seafood cooked. Take them out to drain out the juice.
3. Add 1 ½ T of sesame oil in a pot and add finely chopped garlic, ginger and onion to sauté. Add shredded bacon to stir till golden brown. Add rice to stir for 1 minute with low heat. Drip 1 tablespoon of soy bean sauce to blend evenly. Add juice from Procedure (2) and soup stock (juice and soup stock in 2 cups) and blend well. Add few pinches of salt as seasoning. Cover with lid and turn to medium to low heat to cook till smoking. Then turn to low heat to cook for 25 to 30 minutes (If using a HotPot of better materials, cook about 6 minutes and then turn off the heat. Remove to outer cooker to simmer for another 6 minutes.).
4. Add half the seafood of Procedure (2) into Procedure (3) and blend well with shredded kinchi. Then put the rest half of seafood on top. Drip ½ T of sesame oil. Sprinkle finely-chopped green onion, coriander and white sesame as final garnish.

豐盛的海鮮和酸辣的泡菜拌上以高湯煮的飯是很完美又深具舒爽口感的鍋飯，宴客時或平常的午晚餐超受喜愛。

The rich varieties of seafood, sour and spicy kinchi, with casserole rice cooked with soup stock, make perfect casserole rice bowl with fresh and delicate taste. It is very popular at banquet feast as well as lunch or dinner at any ordinary days.

韓國涼拌墨魚麵

Korean Style Cold Noodle served with Squids

材料：
墨魚（透抽）或花枝 200g、細麵 300g、泡菜 150g（切絲）、小黃瓜 ½ 條（切絲）、紅蘿蔔 ½ 條（切絲）、蔥 1 枝（切絲）、香油 1 大匙

拌料：
韓國辣椒醬 2 大匙、味醂 ⅔ 大匙、香油 1 大匙、醬油 2½ 大匙、蜂蜜 ½ 大匙、韓國辣椒粉適量、胡椒粉少許、蒜泥 ½ 大匙、蔥花 2 大匙

做法：
1. 墨魚切條狀放入鍋內，汆燙熟取出備用。
2. 泡菜切細條狀。
3. 細麵入沸水煮熟取出，沖冷水瀝乾水分，放入調理盆內，倒入 1 大匙香油拌勻。
4. 另取一只調理盆，放入拌料再倒入一半的做法 (1) 拌勻，放入做法 (2) 連同泡菜汁，和做法 (3) 拌勻，放置約 5 分鐘盛入器皿內，擺上小黃瓜絲、紅蘿蔔絲、墨魚及蔥絲。

Ingredient:
Squid 200g, Fine Noodle 300g, Kinchi (Korean Style Pickled Vegetables)(shredded) 150g, Shredded Cucumber ½, Shredded Carrot ½, Shredded Green Onion 1, Sesame Oil 1T

Noodle Sauce:
Korean Chili Paste 2T, Mirin ⅔ T, Sesame Oil 1T, Soy Bean Sauce 2 ½ T, Honey ½ T, Korean Chili Powder few, Pepper Powder few, Garlic Paste ½ T, Finely-chopped Green Onion 2T

Procedure:
1. Cut squid into strips and blanch. Take the blanched squid strips out and leave aside.
2. Shred the kinchi.
3. Cook fine noodles in boiling water and take the cooked noodles out. Cool down with cold water and drain off water. Put the noodles in a basin. Add 1 tablespoon of sesame oil to blend evenly.
4. Take another basin to add ingredients of noodle sauce. Add half of Procedure (1) to mix well. Then add Procedure (2) with kinchi juice. Mix well with Procedure (3). Leave for 5 minutes. Put into a container and add shredded cucumber, carrot, squid and green onion on top.

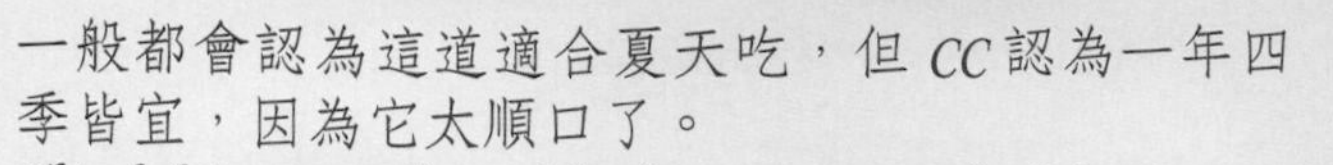

一般都會認為這道適合夏天吃，但 CC 認為一年四季皆宜，因為它太順口了。

This dish is commonly considered as a summer-only dish. However, CC thinks it could be served through the whole year because the delicious taste fits so well with all seasons.

韓式肉醬泡菜拌麵

Korean Style Noodle served with Meat Sauce and Kinchi

材料：
絞肉 900g、蒜末 3 大匙、洋蔥末 ⅓ 杯、泡菜 300g（切絲）、杏鮑菇 2 朵（切絲）、紅蘿蔔適量（切絲）、小黃瓜適量（切絲）、蔥花 1 支、拉麵適量、香菜少許

調味料：
韓國辣椒醬 4 大匙、甜麵醬 2 大匙、味噌 1½ 大匙、醬油 4 大匙、香油 1 大匙、韓國辣椒粉適量

做法：
1. 鍋內入 3 大匙油待熱，放入蒜末爆香，炒至呈微黃色，入洋蔥末以小火炒呈透明狀，即入絞肉炒散。
2. 調味料（香油除外）放入做法 (1) 炒勻，約炒 12 ～ 15 分鐘至顏色較深，最後淋上香油拌勻即成肉醬。
3. 杏鮑菇入鍋煎煮至熟，取出切絲。
4. 煮熟的麵放入器皿內，淋上做法 (2) 擺上泡菜絲、杏鮑菇絲、紅蘿蔔絲、小黃瓜絲，撒上香菜及蔥花。

Ingredient:
Minced Meat 900g, Finely-chopped Garlic 3T, Finely-chopped Onion ⅓ Cup, Kinchi(Korean Style Pickled Vegetables) 300g(shredded), King Oyster Mushroom 2(shredded), Carrot some(shredded), Cucumber some (shredded), Finely-chopped Green Onion 1, Japanese Noodle some, Coriander few

Seasoning:
Korean Chili Paste 4T, Sweet Fermented Flour Paste 2T, Miso 1 ½ T, Soy Bean Sauce 4T, Sesame Oil 1T, Korean Chili Powder few

Procedure:
1. Add 3T of oil in a pot to heat. Add finely chopped garlic to sauté till slightly golden. Add finely chopped onion to stir till transparent (low heat) and then add minced meat to stir till loose.
2. Add ingredients of seasoning (except sesame oil) to Procedure (1) to stir evenly for about 12 to 15 minutes till colored. Finally drip sesame oil and mix well as the meat sauce.
3. Pan-fry and cook king oyster mushroom till cooked and take them out to shred.
4. Add cooked noodles in a container. Drip Procedure (2) and add shredded kinchi, king oyster mushroom, carrot, cucumber on top. Sprinkle coriander and finely chopped green onion as final garnish.

完美烹調寶典 Perfect Cooking Tips

- 可使用牛絞肉。
- 絞肉一定要炒散，才會入味也較不易有腥味。
- 肉醬待涼可放於玻璃罐內，置放冰箱可保存 1 個月，但取用時不可殘留水分。
- 肉醬可夾饅頭或搭配生菜。

- You could also use minced beef.
- Minced meat must be stirred till loose in order to flavor and get rid of flashy smell.
- The meat sauce could be stored in clean glass jar upon cooling down. If stored in a refrigerator, the meat sauce could last for one month. When using it, do not contaminate with any kind of moisture or liquid.
- It is also good with steam buns or served with lettuce.

韓國肉醬跟泡菜的結合是萬人無法擋的，它的魅力相信吃過的人才能體會到。

When the Korean meat sauce and kinchi team up, it is so irresistible. The unique charm can only be understood by those who ate before.

韓國風味牛肉麵
Korean Style Beef Noodle

材料：
排骨 600g、牛五花肉片 300g、馬鈴薯（去皮切塊浸泡水）300g、牛番茄（去皮）2 顆、泡菜 300g、拉麵適量、蔥花適量、香油 1 大匙、蒜頭 3 粒、韓式高湯 2500cc（參見 p.10）、香菜適量

調味料：
韓國辣椒醬 2½ 大匙、黑胡椒粉 1 小匙、鹽適量

醃料：
韓國辣椒醬 1 大匙、蒜末 1 大匙、味噌 1 大匙、醬油 1 大匙、味醂 ½ 大匙、香油 ½ 大匙、黑胡椒粉 ½ 小匙拌匀

做法：
1. 排骨拌入醃料放置 90 分鐘後，放入鍋內煎至呈金黃色取出。
2. 做法 (1) 倒入鍋內注入高湯，燉煮至排骨肉軟（放入壓力鍋待上升二條紅線改小火 8 分鐘）後，再放入牛番茄及馬鈴薯煮至馬鈴薯熟。。
3. 泡菜切絲和泡菜汁倒入做法 (2) 內，入調味料拌匀，入牛五花肉片煮一下至肉熟，即可淋上香油。
4. 煮好的拉麵放入湯碗內，倒入做法 (3) 撒上蔥花及香菜。

Ingredient:
Spare Rib 600g, Beef Rib Finger Slice 300g, Potato (peeled, chucked and soaked in water) 300g, Beef Tomato (peeled) 2, Kinchi (Korean Style Pickled Vegetable) 300g, Japanese Noodle and Finely-chopped Green Onion some, Sesame Oil 1T, Garlic 3, Korean Style Soup Stock 2500cc (please see p.10), Coriander few

Seasoning:
Korean Chili Paste 2 ½ T, Black Pepper Powder 1t, Salt few

Marinate:
Korean Chili Paste 1T, Finely-chopped Garlic 1T, Miso 1T, Soy Bean Sauce 1T, Mirin ½ T, Sesame Oil ½ T, Black Pepper Powder ½ t. Blend all ingredients well.

Procedure:
1. Blend spare ribs with marinate and leave for 90 minutes. Then pan-fry till golden brown and take the pan-fried spare ribs out.
2. Add Procedure (1) in a pot and pour soup stock. Braise till all ribs softened (if using a pressure cooker, turn to low heat and cook for 8 minutes when it rises to two red strips.). Add beef tomato and potato to cook till potato well cooked.
3. Shred the kinchi and use juicer to make kinchi juice. Put shredded kinchi and kinchi juice into Procedure (2). Add seasoning and blend evenly. Then add beef rib finger slices to slightly cook till well done. Drip some sesame oil.
4. Add cooked Japanese noodles in a soup bowl. Pour Procedure (3) and sprinkle finely-chopped green onion and coriander as final garnish.

完美烹調寶典
Perfect Cooking Tips

- 牛肉可使用火鍋肉片。
- 可使用牛腩或牛腱先入高湯約煮至快軟再倒入排骨煮。（如使用壓力鍋煮牛腩，待 10 分鐘後再入排骨、牛腱約煮 16 分鐘）

- Beef rib finger slices could be substituted with hotpot meat slices.
- You could use beef shank or beef tenderloin. First cook with soup stock till almost tendered and then add spare ribs to cook. (If using a pressure cooker to cook beef shank, cook for 10 minutes and then add spare ribs and beef tenderloin to cook another 16 minutes.)

以牛五花肉和燉煮的排骨肉做組合，是一道濃香又帶著微酸微辣的滿分溫暖麵點。
The combination of beef rib finger slices and braised spare ribs together create wonderful heartfelt noodle dish with rich flavor, strong savor and slightly spicy and yet sour taste.

韓式高升元寶湯

Korean Style Dumpling Soup

材料：

水餃 24 ～ 30 個、年糕 500g、火鍋肉片 300g、蔥花適量、紅辣椒末適量、香油 1 大匙、韓式高湯 2500cc（參見 p.10）

醃料：

韓國辣椒醬 ½ 大匙、黑胡椒粉 ½ 小匙、香油 ½ 大匙、蒜泥 ⅓ 大匙、醬油 ⅓ 大匙拌勻

做法：

1. 肉片拌入醃料拌勻置 15 分鐘。
2. 鍋內放入 1 大匙香油，待熱放入做法 (1) 煎熟取出。
3. 高湯注入鍋內待煮沸，放入水餃煮至快熟，倒入年糕煮一下放入適量鹽調味拌勻，盛入器皿內鋪上做法 (2)，撒上蔥花及紅辣椒，淋上適量的香油。
4. 沾醬：韓國辣椒膏 2 大匙、香油 ½ 大匙、醬油 1 大匙、檸檬汁 1 大匙、味醂 1 小匙、蔥花 1 大匙、高湯 1 大匙拌勻

Ingredient:

Dumpling 24 to 30, Rice Cake 500g, Hotpot Meat Slices 150g, Finely-chopped Green Onion and Red Chili Pepper few, Sesame Oil 1T, Korean Style Soup Stock 2500cc (please see p.10)

Marinate:

Korean Chili Paste ½ T, Black Pepper Powder ½ T, Sesame Oil ½ T, Garlic Paste ⅓ T, Soy Bean Sauce ⅓. Blend all ingredients well.

Procedure:

1. Blend hotpot meat slices with marinate and leave for 15 minutes.
2. Add 1T of sesame oil in a pan to heat and add Procedure (1) to pan fry till cooked. Take the pan fried meat slices out.
3. Add soup stock in a pot to boil. Add dumplings to cook till almost cooked and then add rice cakes to slightly cook. Add some salt as seasoning and blend well. Pour into a container with Procedure (2) on top. Sprinkle finely chopped green onion and red chili pepper. Drip few drops of sesame oil.
4. Dip Sauce: Korean Chili Paste 2T, Sesame Oil ½T, Soy Bean Sauce 1T, Lemon Juice 1T, Mirin 1t, Finely-chopped Green Onion 1T, Soup Stock 1T. Blend all ingredients well.

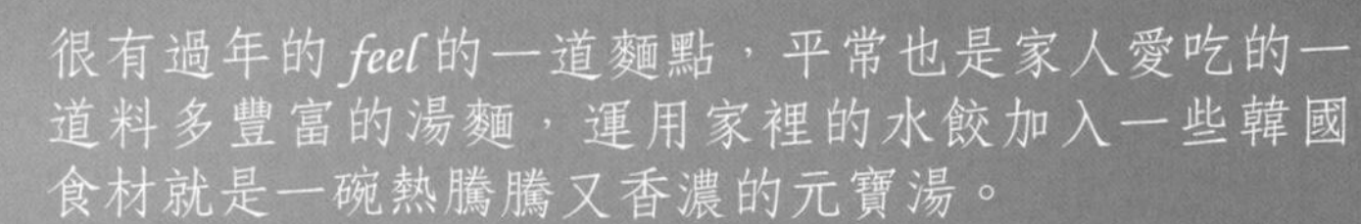

很有過年的 *feel* 的一道麵點，平常也是家人愛吃的一道料多豐富的湯麵，運用家裡的水餃加入一些韓國食材就是一碗熱騰騰又香濃的元寶湯。

This is a flour dish with heavy New Year holiday atmosphere. It is often a rich noodle soup fond by family members. Take good use of dumplings available at home with some Korean food ingredient to create a bowl of hot and savory dumpling soup.

韓國風味什錦鍋

Korean Style Jambalaya

材料：
去骨雞腿 2 隻、烏龍麵 2 包、年糕 1 包、銀芽 100g、綠韭菜 80g、金針菇 1 包、洋蔥 ½ 顆（切絲）、蛤蠣 300g、蝦 150g、泡菜 200g、南瓜 200g、韓式高湯 2500cc（參見韓式高湯 p.10）、蒜末 1 大匙、蔥段 1 段

醃料：
韓國辣椒醬 1 大匙、薑泥 ⅓ 大匙、蒜泥 ½ 大匙、香油 ½ 大匙、醬油 1½ 大匙、味噌 ½ 大匙、味醂 ½ 大匙、蔥末 1 大匙，拌勻

煮料：
韓國辣椒醬 2½ 大匙、洋蔥泥 3 大匙、蒜泥 1½ 大匙、醬油 2½ 大匙、香油 1 大匙、味醂 1 大匙、辣椒粉（韓國）適量、蔥白末適量，拌勻

做法：
1. 泡菜切寬條狀；南瓜切片。
2. 雞腿肉切塊拌入醃料，拌勻放置 30 分鐘後放入鍋內煎熟取出備用。
3. 鍋入 1 大匙油待熱放入蒜末爆香，入蔥段、洋蔥絲炒香，即入年糕略炒幾下注入高湯，依續放入蛤蠣、南瓜、金針菇、烏龍麵、蝦、泡菜（連同泡菜汁）、銀芽和煮料拌勻，最後放上韭菜以適量的鹽調味，並放上做法 (2)。

Ingredient:
Boned Drumstick 2, Udon 2bags, Rice Cake 1 bag, Green Bean Sprouts 100g, Chinese Chive 80g, Noodle Mushroom 1bag, Shredded Onion ½, Clam 300g, Shrimp 150g, Kinchi (Korean Style Pickled Vegetables) 200g, Pumpkin 200g, Korean Style Soup Stock 2500cc (please see p.10), Finely-chopped Garlic 1T, Segmented Leek 1

Marinate:
Korean Chili Paste 1T, Ginger Paste ⅓ T, Garlic Paste ½ T, Sesame Oil ½ T, Soy Bean Sauce 1 ½ T, Miso ½ T, Mirin ½ T, Finely-chopped Green Onion 1T. Blend all ingredients well.

Cooking Sauce
Korean Chili Paste 2 ½ T, Onion Paste 3T, Garlic Paste 1 ½ T, Soy Bean Sauce 2 ½ T, Sesame Oil 1T, Mirin 1T, Chili Powder (Korean) few, Finely-chopped Leek Stalk few. Mix all ingredients evenly.

Procedure:
1. Cut kinchi into wide strips. Slice the pumpkin.
2. Cut drumsticks into chucks and blend with marinate well. Leave for 30 minutes and pan-fry till cooked. Take the cooked drumstick chucks out. Leave aside.
3. Add 1T of oil in a pot to heat and add finely-chopped garlic to sauté. Then add segmented leek and shredded onion to stir till savor. Add rice cakes to stir slightly. Add soup stock. Then add clam, pumpkin, noodle mushroom, udon, shrimps, kinchi (with kinchi juice), green bean sprouts and cooking sauce by the said order to blend evenly. Lastly add Chinese chive and few pinches of salt as seasoning. Add Procedure (2).

完美烹調寶典
Perfect Cooking Tips

- 可做成海鮮鍋不需加麵及烏龍麵，可加冬粉。

- This recipe could be done as seafood hotpot without noodles or udon. You could add green bean fine noodle as preferred.

冬天來上這一鍋暖呼呼的，此鍋學會的同學回家做的機率最高，因為超受家人的喜愛，並且只要料理一鍋可搞定一餐！

In winter classes, students who learn how to cook this dish often cook this at home because it is so fond by all family members. Also just one simple dish could feed up a whole meal. What a deal!

PART 4

令人回味無窮的

中港台料理

荷香糯米飯

Steam Sticky Rice in Lotus Leaf

荷葉經過蒸，它的香氣真是令人受不了，這道糯米飯是CC想到就會做來一解口腹之欲，這幾年更是在端午節時大力推廣，因為不必擔心粽子包不好，又快速，還有就算肉沒包完還可當成一道料理，何樂而不為呢！

Once the lotus leafs are steamed, the smell is so irresisable. This sticky rice dish is often done as CC wishes to satisfy the appetite. In recently years, CC has been largely promoting this dish during Dragon Boast Festival as replacement of Chinese rice Tamale. One reason for this is that we no longer need to worry about not making good Chinese rice tamale. Moreover, it takes very little time to do this dish. Finally, even though there might some meat left, it could still be done as another dish. Why not to do so?

材料：

五花肉1.2kg、長糯米600g、荷葉3大張、鹹蛋黃6個、紅蔥頭末4大匙、乾香菇8朵、蝦米¼杯、杏鮑菇2枝、蔥5枝、蒜頭（去皮）12粒、米酒2大匙、甜麵醬1½大匙、花生300g

調味料：

蠔油1大匙、滷汁1½大匙、醬油1大匙

滷料：

八角2粒、五香粉⅓小匙、老抽6大匙、水2杯、醬油1杯、冰糖3大匙

做法：

1. 花生洗淨泡水約2～3小時後瀝掉水分。
2. 長糯米洗淨浸泡1.8杯水約2小時。
3. 乾香菇浸泡水約6分鐘至香菇軟取出切小丁；蝦米洗淨浸泡米酒約8分鐘取出瀝掉酒；杏鮑菇切丁片狀；五花肉切厚度2.5cm，長10cm。
4. 荷葉洗淨放入沸水汆燙取出再沖冷水，瀝乾水分。
5. 先熱鍋將肉的皮朝鍋底煎至金黃即可取出（只需煎皮）。

Ingredient:

Pork Belly 1200g, Long Sticky Rice 600g, Lotus Leaf 3, Salted Egg Yolk 6, Finely-chopped Shallot 4T, Dried Shitake Mushroom 8, Shelled Dried Small Shrimp ¼ Cup, King Oyster Mushroom 2, Leek 5, Garlic (peeled) 12, Rice Wine 2T, Sweet Fermented Flour Paste 1 ½ T, Peanut 300g

Seasoning:

Oyster Sauce 1T, Braise Sauce 1 ½ T, Soy Bean Sauce 1T

Braise Sauce

Aniseed 2, Five Spice Powder ⅓ T, Dark Soy Bean Sacue 6T, Water 2Cups, Soy Bean Sauce 1 Cup, Crystal Sugar 3T

Procedure:

1. Rinse peanuts and soak in water for 2 to 3 hours. Drain off water.
2. Rinse long sticky rice and soak in 1.8 cup of water for 2 hours.
3. Soak dried shitake mushroom for 6 minutes till soften. Take the soaked shitake mushroom and cut into dices. Rinse the shelled dried small shrimps and soak in rice wine for 8 minutes. Take the soak shrimps out and drain off wine. Slice king oyster mushroom and then cut into dices. Cut pork belly into strips that are 2.5cm thick and 10cm long.
4. Rinse and blanch lotus leafs in boiling water. Take them out and cool down with running cold water. Drain off water.
5. Heat a pan and put the skin side of pork belly facing the pan to fry till golden and then take them out (only the skin part needs to be pan fried).

6. 做法 (5) 的鍋內放入蒜頭、蔥（整枝），爆香至呈微焦黃色，再倒入甜麵醬淋上米酒（小火）炒至香味溢出，倒入煎好的五花肉和花生放進滷料，燉滷約 1 ～ 1.5 小時至肉軟Q（使用壓力鍋，待上升二條線改小火約 22 分鐘，鍋內不需放水，老抽 2 大匙、醬油 2½ 大匙、冰糖 2 大匙）後取 2 大匙）後取出一塊肉切成 3 塊。
7. 糯米連同浸泡的水入鍋內（或電鍋）煮熟。
8. 鍋內倒入 2 大匙油放入紅蔥頭爆香至呈微焦黃色，倒入蝦米炒香再放入香菇丁爆香，倒入杏鮑菇，撒上白胡椒粉少許，倒進煮好的糯米飯放入調味料及 ⅓ 小匙的五香粉，拌炒均勻。
9. 將荷葉剪掉蒂頭，一張剪成 3 張，先取一張放入做法 (8)，舖上做法 (6) 再放進蛋黃包成捲筒狀，放入蒸籠內蒸約 40 分鐘，壓力鍋蒸約 15 分鐘。

6. Add garlic and leeks (in whole piece) in the pot of Procedure (5) to sauté till slightly golden brown. Then add sweet fermented flour paste and rice wine (low heat) to stir till savor. Add pan-fried pork and peanut along with braise sauce. Brasie for 1 to 1.5 hours till pork tendered (If using a pressure cooker, turn to low heat to cook about 22 minutes when it rises to two (No water is needed, add dark soy bean sauce 2T, soy bean sauce 2 ½ T, crystal sugar 2T). Take the meat out and cut into three pieces.
7. Cook the sticky rice together with the soaking water into a pot (or electronic rice cooker) till well done.
8. Add 2T of oil in a pan and add shallots to sauté till slightly golden brown. Add dried shrimps and shitake mushroom dices to sauté. Then add king oyster mushroom. Sprinkle few pinches of white pepper poweder. Pour cooked sticky rice, along with seasoning and ⅓ t of five spice powder to blend well.
9. Cut off the stalk of lotus leafs. Trim each one leaf into 3 small ones. Take one and put Procedure (8) inside, layer with Procedure (6) with egg yolk, and wrap it into a cone. Steam those cones in steaming cage for 40 minutes. If using a pressure cooker, it takes about 15 minutes.

完美烹調寶典
Perfect Cooking Tips

· 蝦米浸泡米酒可去腥增加香味。
· 荷葉入沸水氽燙殺菌亦使荷葉軟化。
· 蒜及蔥一定要爆炒至呈焦黃色滷的肉才會香。

· Soak dried shrimps in rice wine could reduce the fish smell as well as increase the savor.
· Blanch those lotus leafs in boiling water in order to kill germs and soften those leafs as well.
· Garlic and leek must be sauté till nicely golden brown so the meat braised with them will be strongly savored.

客家粿條蒸鮮蝦

Steam Hakka Flat Rice Noodles with Fresh Shrimps

材料：
粿條 600g、蝦 10 隻、香菜末 1 大匙、蔥末 1 大匙、紅辣椒末少許、蒜酥 2½ 大匙

調味料：
高湯 ½ 杯、蒜末 1 大匙、蔥末 1 大匙、香菜末 1 大匙、嫩薑末 1 大匙、紅辣椒末適量、米酒 ½ 大匙、香油 1 大匙、白胡椒粉 ½ 小匙、白砂糖 ⅓ 小匙、醬油 3 大匙拌勻。

蒜酥：
油 1 杯倒入小鍋內待熱倒入蒜末 ¼ 杯，炸至呈微黃即馬上撈出，如已呈金黃色再撈出會苦，因為油溫很高。

做法：
1. 蝦由背部劃開取出腸泥，攤開蝦身段筋（以防捲曲）。
2. 取一深盤倒入粿條舖上做法(1)，淋上調味料，撒上 1½ 匙的蒜酥放入蒸籠，蒸約 6 ~ 8 分鐘（使用休閒鍋，蓋上過蓋待冒煙改小火煮 3 分鐘）。
3. 做法 (2) 取出撒上香菜末、蔥末、紅辣椒及剩餘 1 大匙蒜酥。

Ingredient:
Flat Rice Noodle 600g, Shrimp 10, Finely-chopped Coriander 1T, Finely-chopped Green Onion 1T, Finely-chopped Red Chilly Pepper few, Deep Fried Garlic 2 ½ T

Seasoning:
Soup Stock ½ cup, Finely-chopped Garlic 1T, Finely-chopped Green Onion 1T, Finely-chopped Coriander 1T, Finely-chopped Fresh Ginger 1T, Finely-chopped Red Chilly Pepper few, Rice Wine ½ T, Sesame Oil 1T, White Pepper Powder ½ T, Fine White Sugar ⅓ T, Soy Bean Sauce 3T. Blend all ingredients well.

Deep Fried Garlic
Add 1 cup of oil in a small pot to heat and add ¼ cup of finely chopped garlic to deep fry till slightly golden. Take the fried garlic out right away. If it turns golden brown before taking out, it will taste bitter since the oil temperature is very high.

Procedure:
1. Cut the back of shrimps to remove the intestine mud. Press the shrimps flat and cut off tendon (so that shrimps won't shrink during cooking).
2. Take one deep plate and pour flat rice noodles onto it. Put Procedure (1) on top of rice noodles. Drip seasoning and sprinkle 1 ½ T of deep fried garlic. Put into steaming cage to steam for 6 to 8 minutes (If using HotPan: steam with pot cover still smoking and turn to low heat for another 3 minutes).
3. Take Procedure (3) out and sprinkle finely chopped coriander, green onion, red chilly pepper and the rest 1T of deep fried garlic as final garnish.

・亦可使用去骨的魚肉代替例如鱈魚、鲂魚、鯛魚、鰈魚。

・You could substitute with boned fish filet such as codfish, Dory fish, snapper or flatfish in this receipe.

CC的學生們強力推薦一定要學的一到菜！簡單又豐富。

All CC's students highly recommend this dish as a must-learn one! It is so easy to do and so fancy and delicious to enjoy.

沙茶魷魚羹米粉

Satay Sauce Squid Potage Soup with Rice Vermicelli

CC太喜歡沙茶魷魚羹了，所以才會做出自己喜歡的口味，這次想跟大家分享CC的私房味道。

CC is so much in love with this dish that CC decides to create one's own flavor. In this book CC would like to share this personal recipe with all readers.

材料：
乾魷魚 150g、魚漿 400g、沙茶醬適量、沙拉筍 120g（切絲）、白蘿蔔 200g（切小丁塊狀）、九層塔適量、太白粉 4 大匙、紅蔥頭酥 2 大匙、蒜泥適量、柴魚 1 大把

高湯：
洋蔥 1 個（切塊）、雞骨 600g、豬骨 600g、昆布 1 大塊、水 2000cc，放入鍋內熬煮 5 小時後過濾（若使用壓力鍋，熬煮 1 小時）。

調味料：
白胡椒粉適量、冰糖 ½ 匙、黑醋 1½ 大匙、白醋 ½ 大匙、香油 1 大匙、鹽適量

做法：
1. 魷魚洗淨，浸泡水約 1 小時後取出，切條狀倒入 1 大匙太白粉拌勻。
2. 魚漿倒入調理盆內，放入紅蔥頭酥拌勻。
3. 將做法 (1) 的魷魚裹上做法 (2) 放入高湯中煮至浮出水面，撈出即成魷魚羹。
4. 白蘿蔔和竹筍絲放入做法 (3) 的高湯內，煮至白蘿蔔熟透再倒入魷魚羹，並放入調味料及柴魚拌勻，最後以太白粉水勾芡淋上香油。
5. 煮熟的米粉放入碗內倒入做法 (4)，放入適量的沙茶醬和少許的蒜泥，撒上九層塔。

Ingredient:
Dried Squid 150g, Fish Paste 400g, Satay Sauce few, Processed Bamboo Shoot for Salad use 120g(shredded), Daikon Radish 200g(dice), Basil Leaf few, Potato Strach 4T, Deep-fried Shallet 2T, Garlic Paste few, Dried Stockfish Flake 1 large pinch

Soup Stock
Onion 1(chucked), Chicken Bone 600g, Pork Bone 600g, Kelp 1 large piece, Water 2 liter

*Put all ingredients in a pot to barise for 5 hours and then filter (If using a pressure cooker, braise for 1 hour.)

Seasoning:
White Pepper Powder few, Crystal Sugar ½ T, Black Vinegar 1½T, White Vinegar ½ T, Sesame Oil 1T, Salt few

Procedure:
1. Rinse the dried squid and soak in water for 1 hour. Take the processed squid out and cut into stirps. Blend with 1T of potato starch.
2. Add fish paste in a basin and blend well with deep fried shallots.
3. Wrap squid strips of Procedure (1) with Procedure (2) and cook in soup stock till the wrapped squid strips floating over. Take those strips out.
4. Add daikon radish and shredded bamboo shoots into the soup stock of Procedure (3) to braise till daikon radish well cooked. Then add wrapped squid strips, and blend well with seasoning and dried stockfish flakes. Thicken with potato starch water and sprinkle sesame oil.
5. Put cooked rice vermicelli in a bowl and pour Procedure (4). Add few satay sauce and garlic paste. Sprinkle few basil leafs before serve.

完美烹調寶典
Perfect Cooking Tips

- 生筍帶殼放入洗米水煮至筍熟，去殼切絲。
- 魷魚拌上太白粉增強其黏著性。
- 紅蔥頭作法見越南咖哩什錦河粉（參見 p.72）做法 (1)。

- Cook raw bamboo shoot with outer skin in rice-rinsing water till well done. Then peel and shred the cooked bamboo shoot.
- Blend squid with potato starch to increase the sticky extent.
- For the procedure of deep fried shallot please see the procedure (1) of Vietnam Curry Pho (see p.72).

川味海鮮鍋麵

Sichuan Style Seafood Noodle Soup

材料：

魚 1 條、牡蠣 150g、蝦 200g、透抽 1 隻、蛤蜊 200g、貽貝 8 個、大白菜 300g、麵適量、蒜頭 10 粒、辣豆瓣醬 1½ 大匙、高湯 2000cc、乾辣椒 5 枝、紅辣椒 2 枝、米酒 1½ 大匙、蒜苗 1 枝（切段）、番茄醬 2 大匙、香菜少許

高湯：

雞骨 900g、蔥 3 枝、薑 3 片、花椒粒 1 大匙、八角 2 粒、小魚乾 ⅓ 杯、水 2500cc，熬煮約 4 ～ 5 小時，使用壓力鍋，待上升二條紅線改小火 1 小時後過濾。

做法：

1. 魚洗淨擦乾水分撒上少許鹽，牡蠣以麵粉（2 大匙）、酒（1 大匙）輕輕拌洗，再以小水注沖洗淨；蝦去殼留尾巴；透抽切圈狀。
2. 鍋內放入 1 大匙油待熱放入蒜、乾辣椒、紅辣椒爆香後，即入辣豆瓣醬以中小火炒至香味溢出，淋上米酒炒香並放入番茄醬，注入高湯煮沸放進大白菜煮軟。
3. 魚入鍋煎熟取出備用；麵入沸水煮熟取出置器皿內。
4. 將海鮮放入做法 (2) 內煮熟熄火，放入煮好的麵鋪上煎好的魚，撒上蒜苗段、香菜末。

Ingredient:

Fish 1, Oyster 150g, Shrimp 200g, Squid 1, Clam 200g, Mussel 8, Chinese Cabbage 300g, Noodle few, Garlic 10, Sticky Chili Bean Sauce 1 ½ T, Soup Stock 2000cc, Dried Chilly Pepepr 5, Red Chilly Pepper 2, Rice Wine 1 ½ T, Garlic Bolt 1 (segmented), Ketchup 2T, Coriander few

Soup Stock

Chicken Bone 900g, Leek 3, Ginge 3 slices, Sichuan Pepper 1T, Aniseed 2, Dried Samll Fish ⅓ Cup, Water 2500cc. Braise all ingredients together for 4 to 5 hours or cook with pressure cooker, turn to low heat to cook for 1 hour when it rises to two red strips. Then filter.

Procedure:

1. Rinse fish and wipe dry. Sprinkle salt. Gentily rinse oyster with flour (2T) and wine and then rinse carefully with very gentile water column. Shell the shrimps with tails left. Cut the squid into rings.
2. Add 1T of oil to heat and add garlic, dried chilly pepper, and red chilly pepper to sauté. Add sticky chili bean sauce (turn to medium to low heat) to stir till savor. Drip rice wine to stir till savor. Add ketchup and soup stock to cook till boil. Then add Chinese cabbage to cook till tendered.
3. Pan fry the fish and take it out. Put noodles into boiling water till well cooked. Place cooked noodle in a container.
4. Put processed seafood into Procedure (2) to cook till well done. Turn off the heat. Add cooked noodle and layer the pan-fried fish on top. Sprinkle segmented garlic bolt and finely chopped coriander to serve.

完美烹調寶典 Perfect Cooking Tips

- 洗牡蠣時不可用力搓洗，水注也不宜過大，以免洗破牡蠣。
- 此高湯可應用於煮粥，或川味火鍋鍋底均可。

- Do not rinse oyster too harsh and the water column should be gentiel otherwise it is very easy to damage those oysters.
- This soup stock is ideal for congee or as hotpot base for Sichuan style hotpot.

非常過癮的麵點，豐富的海鮮料配上口感鮮美的湯頭，總是令人回味，亦可當川味海鮮鍋，又可當麵點是個多運用的料理。

This is a very fancy noodle dish. Aboundant and assorted seafood with delicious soup base is always so unforgettably tasty. This dish could be served as Sichuan style seafood hotpot as well as a noodle dish.

港式豉汁排骨煲仔飯

Hong Kong Style Black Bean Sauce Spareribs Casserole Rice

材料：

小排骨 600g（剁小塊狀）、豆豉 2½ 大匙、蒜末 1½ 大匙、米酒 2 大匙、紅辣椒 1 條（切末）、青江菜 200g、米 2 杯、西式高湯 2 杯（參見 p.9）

醃料：

蒜末 1 大匙、酒 1 大匙、醬油 1 大匙、醬油膏 1 大匙、白砂糖 1 小匙、白胡椒粉 ½ 小匙、蒜油 1 大匙、太白粉 1½ 大匙

做法：

1. 豆豉置於碗內倒入米酒浸泡 10 分鐘後剁碎。
2. 小排骨倒入醃料及做法 (1) 拌勻，放置 60 分鐘後放入蒸籠蒸約 40 分鐘（使用壓力鍋，待上升二條紅線，以小火 12 分鐘）。
3. 鍋中放入 1 大匙油，倒入蒜末 1 大匙爆香至微黃，放入米略炒注入高湯，蓋上鍋蓋待冒煙改小火煮約 25 ～ 20 分鐘（使用休閒鍋，冒煙煮 6 分鐘，移外鍋燜 6 分鐘）。
4. 鍋入 1 大匙油待熱放入 ½ 大匙蒜末爆香，倒入青江菜及適量的鹽炒熟取出。
5. 取砂鍋放於爐上加熱倒入做法 (3)，鋪上做法 (2)，擺上做法 (4) 即可取出置盤中。

Ingredient:

Spare Ribs(chop into small chucks)600g, Marinate:d Black Bean 2 ½ T, Finely-chopped Garlic 1 ½ T, Rice Wine 2T, Finely-chopped Red Chilly Pepper 1, Bok Coy 200g, Rice 2 cups, Western Style Soup Stock 2 cups (please see p.9)

Marinate:

Finely-chopped 1T, Wine 1T, Soy Bean Sauce 1T, Thick Soy Bean Sauce 1T, Fine White Sugar 1T, White Pepper Powder ½ T, Garlic Oil 1T, Potato Starch 1 ½ T

Procedure:

1. Soak marinated black beans with rice wine for 10 mintues and finely chop.
2. Blend spare ribs with marinate and Procedure (1) and leave for 60 minutes. Then steam in steaming cage for 40 minutes (If using a pressure cooker, you should turn to low heat to cook for 12 minutes when it rises to two red strips.)
3. Add 1T of oil with finely chopped garlic to sauté till slightly yellowish. Add rice to stir and then pour soup stock to cook with cover till smoking. Turn to low heat to braise for 25 to 20 minutes (If cooking with HotPan, cook 6 minutes upon smoking and then remove to simmer for another 6 minutes).
4. Add 1T of oil to heat and then add ½ T to sauté. Add bok coy and salt to stir till cooked. Take them out.
5. Put casserole on the stove to heat and add Procedure (3), on top layer with Procedure (2) and granish with Procedure (4). It is ready to serve with plate.

在港式茶餐廳，這道料理是很多人喜歡點的一道煲仔飯，但有時吃到的口味並不足以讓 CC 有幸福的感覺，所以 CC 想介紹給大家讓 CC 吃得有幸福感的做法。

At Hong Kong style tea houses this dish is a highly popular choice of casserole rice among costumers. Yet sometimes the taste is not good enough to satisfy CC. Therefore CC would like to share the recipe which pleases CC well.

田園野菇雞丁飯

Garden Style Assorted Mushroom and Chicken Dice on Rice

材料：

去骨雞腿 3 隻、鮮香菇 5 朵、杏鮑菇 2 朵、蘑菇 100g、香菜梗末 1 大匙、蒜末 1 大匙、洋蔥末 3 大匙、米 2 杯、西式高湯 1 杯又 9 分滿（參見 p.9）、香菜葉少許

調味料：

醬油 1 大匙、鹽適量、白胡椒粉少許

醃料：

醬油 2½ 大匙、酒 ½ 大匙、香油 ½ 大匙、白胡椒粉 ½ 小匙、蒜末 1 大匙、香菜末 1 大匙，拌勻

做法：

1. 雞腿肉切塊拌入醃料置 30 分鐘。
2. 鮮香菇切長條狀，杏鮑菇切丁塊狀，蘑菇切片狀。
3. 鍋內放入 ½ 大匙油待熱放入做法 (1)，皮先朝鍋底煎再翻面煎至呈焦黃色，取出備用；蒜末倒入爆香，再入香菜梗末及洋蔥末炒香，倒入做法 (2) 略炒幾下，即入米炒約 1 分鐘（小火），放入調味料拌勻；放入煎好的雞塊及高湯拌勻，改中小火，蓋上鍋蓋待冒煙改小火，一般鍋約煮 25～30 分鐘（使用休閒鍋，待冒煙改小火 6 分鐘，移入外鍋燜 6 分鐘）。
4. 盛入器皿內撒上香菜葉。

Ingredient:

Boned Drumstick 3, Fresh Shitake Mushroom 5, King Oyster Mushroom 2, Mushroom 100g, Finely-chopped Coriander Stem 1T, Finely-chopped Garlic 1T, Finely-chopped Onion 3T, Rice 2 Cups, Western Style Soup Stock 1.9 Cups (please see p.9), Coriander Leaf few

Seasoning:

Soy Bean Sauce 1T, Salf and White Pepper Powder few

Marinate:

Soy Bean Sauce 2 ½ T, Wine ½ T, Sesame Oil ½ T, White Pepper Powder ½ T, Finely-chopped 1T, Finely-chopped Coriander 1T. Blend all well.

Procedure:

1. Chop drumsticks into chucks and blend with marinate. Leave for 30 minutes.
2. Cut fresh shitake mushroom into long strips, king oyster mushroom into dices, and mushroom into slices.
3. Add ½ T of oil in a pan to heat and then add Procedure (1). Put the skin side of drumsticks facing down to pan fry till golden brown and then put the other side against the pan till golden brown as well. Take drumsticks out. Add finely chopped garlic to sauté and then add finely chopped coriander stems and onion to stir till savor. Add Procedure (2) to slightly stir and then add rice to stir for 1 minute (low heat). Add seasoning and blend well. For general type of pots, it requires 25 to 30 minutes to cook (If cooking with HotPan, cook 6 minutes upon smoking and then remove to simmer for another 6 minutes).
4. Pour into a container and sprinkle some coriander leafs as final garnish.

完美烹調寶典 Perfect Cooking Tips

・蘑菇切開亦發黑，可拌入檸檬汁或要炒時再快速切片。

・Mushroom is easily darken upon cutting. You could blend it with lemon juice or cut right before ready to stir in pan.

很豐盛的一道菜飯，最受媽媽們及帶便當族群的喜愛，因為只要做一道就可準備好便當菜了。

This is a very fancy vegetable on rice dish. It is quite fond by housewives and lunchbox eaters due to the fact that one simple dish contains everything in lunchbox.

紅燒牛腩煲飯

Braised Beef Brisket Casserole Rice

很多 CC 的學生和朋友，很喜歡這道料理，因為這道料理有蔬菜又有肉，一鍋就可以應付一餐了，不僅家人喜愛又可宴客用。

Many CC's students as well as friends like to cook this dish very much. Since this dish provides both vegetables and meats as a meal. Famiy members all enjoy this dish while it could be served at banquest for guests.

材料：
牛肋條（牛腩）1kg、芥蘭菜 300g、紅蘿蔔 1 小條、洋蔥 1 個（切絲）、蔥花少許、白飯適量

調味料 A：
醬油 3 大匙、醬油膏 2 大匙、蔥段 2 段、薑片 3 片、八角 2 粒、桂皮 1 枝、花椒粒 1 大匙、酒 1 大匙、高湯 2 杯

調味料 B：
番茄醬 3 大匙、白胡椒粉 1 小匙、醬油膏 1½ 大匙、酒 1 大匙、白砂糖 1 小匙

做法：
1. 牛肋條切塊放入鍋內煎至呈金黃色取出備用。紅蘿蔔削皮切滾刀塊狀。
2. 做法 (1) 的鍋入 1 大匙油待熱，放入蔥段、薑片爆香，倒入做法 (1) 灑上米酒，放入調味料 A，燉煮至肉快軟約 50 分鐘後放入紅蘿蔔，煮至紅蘿蔔熟（使用壓力鍋上升紅線改小火煮 18 分鐘，再放入紅蘿蔔上升 2 條紅線改小火煮 2 分鐘），後將湯汁過濾。
3. 鍋入 ½ 大匙油倒入洋蔥絲爆香，入調味料 B 及做法 (2) 的湯汁及牛肉再燉煮約 10 分鐘（使用壓力鍋待上升二條紅線以小火約 2 分鐘）。
4. 芥蘭菜氽燙熟取出。
5. 氽燙好的芥蘭菜放入砂鍋底倒入做法 (3) 煮一下，撒上蔥花即可，再搭配一碗白飯。

Ingredient:
Beef Brisket 1kg, Chinese Kale 300g, Carrot 1, Onion 1 (shredded), Finely-chopped Green Onion few, Steamed Rice some

Seasoning A:
Soy Bean Sauce 3T, Thicken Soy Bean Sauce 2T, Leek Segments 2, Ginger 3 slices, Aniseed 2, Chinese Cinnamon 1, Sichuan Pepper 1T, Wine 1T, Soup Stock 2 Cups

Seasoning B:
Ketchup 3T, White Pepper Powder 1T, Thicken Soy Bean Sauce 1 ½ T, Wine 1T, Fine White Sugar 1T

Procedure:
1. Cut beef briskets into chucks and pan fry in a pan till golden brown. Take them out and leave aside. Peel carrot and cut in round chucks.
2. Add 1 tablespoon of oil in the pan of Proccedure (1) to heat and add leek segments and ginger slices to sauté. Then add Procedure (1) and sprinkle rice wine along with seasoning A to braise 50 minutes till meat tendered. Add carrot to cook till well cooked (If using a pressure cooker, turn to low heat to cook 18 minutes when it rises to two red strips. Then add carrot to cook till rising to two red strips and turn to low heat for another 2 minutes), then filte the sauce.
3. Add ½ tablespoon of oil in a pan and add shredded onion to sauté with seasoning B and sauce from Procedure (2) with beef to braise for another 10 minutes (If using a pressure cooker, turn to low heat to cook for another 2 minutes when it rises to two red strips).
4. Blanch Chinese kale and take them out.
5. Put the blanched Chinese kale on the bottom of casserole and add Procedure (3) to cook a little bit. Sprinkle finely-chopped green onion as final garnish. Serve with a bowl of steamed rice.

完美烹調寶典
Perfect Cooking Tips

- 可使用牛腱或豬梅花肉或豬腱肉。
- 使用壓力鍋不需放高湯，醬油或醬油膏可少一倍的量。

- You could substitute with beef shank, pork shoulder or pork shank.
- When using pressure cooker, no soup stock is needed and the use of soy bean sauceor thicken soy bean sauce could be thus reduced to half.

港式蝦丸皮蛋粥

Hong Kong Style Congee with Shimp Balls and Thousand Years Eggs

港式的粥品中，CC最愛這道了，它就是有說不出的好滋味！

Among the Hong Kong style congees, this dish remains CC's favorite one because of its speechlessly good taste!

材料：
皮蛋 3 個、米 1 杯、嫩薑絲、蔥花、香菜末各適量

蝦丸：
蝦仁剁碎 300g、花枝漿 150g、荸薺 6 粒（拍碎切末去掉汁）、蔥末 1 大匙、香菜末 1 大匙、白胡椒粉 ½ 小匙、鹽 ⅓ 小匙、蛋 ½ 個、玉米粉 2 大匙，一起放入調理盆內拌勻至有黏性。

高湯：
蝦殼 300g ＋雞骨 600g ＋豬骨 600g ＋蔥 2 枝＋薑 3 片＋水 2500cc，放入鍋內熬煮 4 ～ 5 小時（使用壓力鍋，待上升二條紅線，以小火煮 50 分鐘）後過濾

做法：
1. 高湯 9½ 杯倒入湯鍋內（小火）煮。
2. 蝦丸做成圓球狀放入做法 (1) 的高湯內煮熟取出。
3. 米倒入鍋內小火炒至呈象牙白的色澤，倒入做法 (2) 高湯煮至呈稠狀（使用壓力鍋，待上升二條紅線，以小火 6 分鐘），以適量的鹽和白胡椒粉調味。
4. 皮蛋去殼切塊倒入做法 (3)，並將蝦丸倒入拌勻。
5. 將做法 (4) 盛於器皿內灑上蔥花、香菜末、薑絲。

Ingredient:
Thousand Years Egg 3, Rice 1 cup, Shredded Fresh Gingers, Finely-chopped Green Onion and Coriander few for each

Shrimp Ball
Finely-chopped Shelled Shirmp 300g, Squid Paste 150g, Waterchestnut 6(crush, chop finely and drain dry), Finely-chopped Green Onion 1T, Finely-chopped Coriander 1T, White Pepper Powder ½ T, Salt ½ T, Egg ½, Cornstarch 2T. Blend all said ingredients well in a basin till sticky.

Soup Stock
Shrimp Shell 300g, Chicken Bone 600g, Pork Bone 600g, Leek 2, Ginger 3 piece, Water 2500cc. Braise in a pot for 4 to 5 hours (If using a pressure cooker, you should turn to low heat to cook for 50 minutes when it rises to two red strips) and then filter.

Procedure:
1. Add 9 ½ cups of soup stock in a pot to braise with low heat.
2. Shape shrimp ball mixure into balls and put into soup stock from Procedure (1) to cook till well down. Take out the cooked shrimp balls.
3. Add rice into pot to stir with low heat till rice turning to ivory white color. Add the soup stock from Procedure (2) to braise till thickened (If using a pressure cooker, you should turn to low heat to cook for 6 minutes when it rises to two red strips.) Add with salt and white pepper powder for seasoning.
4. Shell those thousand-year eggs and chop into chucks. Add Procedure (3) and shrimp balls to mix well.
5. Pour Procedure (4) into a container and sprinkle with finely-chopped green onion, coriander and shredded ginger as final .

完美烹調寶典
Perfect Cooking Tips

- 米炒過才會香。
- 如喜歡粥稀一些可多放些高湯。
- 高湯可做成鍋底煮麵等。
- 煮蝦丸時要小火，蝦丸才不易破裂。

- Rice is suggested to stir for savor.
- If you prefer more liguid for your congee, you could add extra soup stock.
- This soup stock could be soup base for noodle.
- Cook shrimp balls with low heat, shrimp balls is not easily broken.

扁魚筍丁芋香粥

Hong Kong Style Congee with Shimp Balls and Thousand Years Eggs

材料：

雞胸肉 300g（去皮去骨）、扁魚 6 片、芋頭 600g、綠竹筍 2 枝或沙拉筍 1 包、米 2 杯、西式高湯 2300cc（參見 p.9）、芹菜末適量

調味料：

醬油 2 大匙、白胡椒粉 1 小匙、鹽適量

做法：

1. 扁魚放入烤箱以上下火 160℃烤至酥脆取出剁碎，芋頭削皮切丁狀，雞胸肉剁泥狀。
2. 綠竹筍連同外殼放入鍋內，倒入洗米水煮至熟，取出剝去外殼切丁。
3. 鍋中倒入米（不放油）小火炒約 1 分鐘，注入高湯、筍丁，並倒入芋頭約煮 25 分鐘呈粥狀。（使用壓力鍋，待上升二條紅線，以小火煮 6 分鐘）放入扁魚約煮 3 分鐘，再倒入雞胸肉末，放進調味料拌勻，盛入器皿內撒上芹菜末。

Ingredient:

Chicken Breast 300g(skinned and bonned), Dried Sliver Carp 6 pieces, Taro 600g, Green Bamboo Shoot 2 or Processed Salad-use Bamboo Shoot 1 bag, Rice 2 cup, Western Style Soup Stock 2300cc (please see p.9), Finely-chopped Salary few

Seasoning:

Soy Bean Sauce 2T, White Pepper Powder 1T, Salt few

Procedure:

1. Roast dried silver carp in an oven with top and bottom heat of 160℃ till crispy. Take the roasted silver carp out and torn into pieces. Skin taro and chop into dices. Chop chicken breast into paste.
2. Put green bamboo shoot with outer skin together in a pot. Pour rice-rinsing water to cook till well done. Take the cooked bamboo shoots out, peel and chop into dices.
3. Add rice in a pot (without oil) and stir about 1 minute. Then add soup stock, bamboo shoot dices, and taro to cook for another 25 minutes into congee. (If using a pressure cooker, you should turn to low heat to cook for 6 minutes when it rises to two red strips.) Add dried sliver carp to cook for 3 minutes. Then add finely-chopped chicken breast and seasoning to blend well. Pour into a container and sprinkle finely chopped salary as granish.

完美烹調寶典 Perfect Cooking Tips

· 扁魚亦可放入油鍋炸至酥脆。
· 米炒過才會更香。

· You could also deep fry the dried silver carp in a pot till golden crispy.
· Rice is suggested to stir for savor.

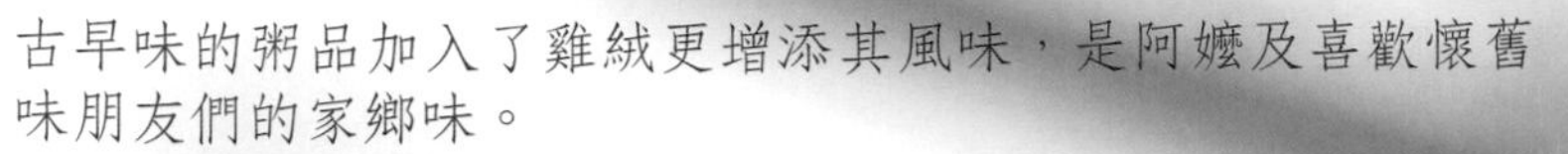

古早味的粥品加入了雞絨更增添其風味，是阿嬤及喜歡懷舊味朋友們的家鄉味。

Congee with Dried Silver Carp, Bamboo Shoot Dice and Taro

This old-fashion style congee with finely chopped chicken has rich flavor and it is such homy flavor for grandmas and those who are fond of nostalgia taste.

山藥鮮魚粥

Chinese Yam Congee with Fish Filet

材料：
米 2 杯、西式高湯 18 杯（1 杯米對 9 杯高湯）（參見 p.9）、山藥 600g、魚肉 600g（去皮去骨、切塊）、嫩薑絲適量、蔥花適量、嫩薑 2 片

調味料：
白胡椒粉 1 小匙、鹽適量

醃料：
蔥 2 枝（拍扁）、薑 3 片（拍扁）、米酒 1 大匙、香油 ½ 匙、鹽 ½ 小匙

做法：
1. 先將蔥、薑與米酒一起抓勻，抓擠出蔥、薑汁，再倒入魚肉、鹽及香油拌勻置 30 分鐘。
2. 鍋中倒入米（小火）炒至呈象牙白注入高湯，並放進 2 片嫩薑熬煮成粥（使用壓力鍋，待上升二條紅線，以小火煮 6 分鐘）。
3. 山藥削好皮切成滾刀塊狀後放入做法 (2) 內約煮 3 分鐘，放入調味料拌勻，再倒入做法 (1) 煮至魚熟即可。
4. 做法 (3) 盛入器皿內撒上蔥花，擺上薑絲。

Ingredient:
Rice 2 cups, Western Style Soup Stock 18 cups (1cup of rice per 9 cups of soup stock) (please see p.9), Chinese Yam 600g, Skinned and Boned Fish Filet 600g(chop in chucks), Shredded Fresh Ginger and Finely-chopped Green Onion few, Fresh Ginger 2 slices

Seasoning:
White Pepper Powder 1T, Salt few

Marinate:
Leek 2(crush flat), Ginger 3 slices (crush flat), Rice Wine 1T, Sesame Oil ½ T, Salt ½ T

Procedure:
1. Blend leek, ginger and rice wine well and squeeze juice. Then add fish filet, salt and sesame oil to mix evenly. Leave for 30 minutes.
2. Add rice in a pot to stir with low heat till turning ivory white. Pour soup stock with 2 slices of fresh ginger to braise into congee. (If using a pressure cooker, you should turn to low heat to cook for 6 minutes when it rises to two red strips.)
3. Add peeled and chucked Chinese yam into Procedure (2) to cook for 3 minutes. Add seasoning to blend well. Add Procedure (1) to cook till fish filet well cooked.
4. Pour Procedure (3) in a container and sprinkle with finely chopped green onion and shredded ginger as final granish.

・米以乾鍋（不放油）炒過才會香。
・魚片調好味道時再放入鍋內，以防魚片在拌攪時破裂。

・Rice is suggested to stir in a dry pan wihout oil in order to savor.
・Fish filet should be added after seasoning done so that it won't be torn apart during blending.

健康又美味的粥，是 CC 常煮給大家吃的粥品，簡單又受歡迎，尤其山藥更添加了粥的口感，增加養生功效。

This is a very nutrious and tasteful congee. CC often cooks this for friends. It is simple and easy, and very popular among friends. In particular the Chinese yam adds great texture for this congee dish plus function of wellness being.

異國風
主食料理

焗烤、燉飯、粥品、鍋物等
60道美味幸福上桌

作　　者　洪白陽
攝　　影　楊志雄

發 行 人　程安琪
總 策 畫　程顯灝
編輯顧問　錢嘉琪
編輯顧問　潘秉新

總 編 輯　呂增娣
主　　編　李瓊絲　鍾若琦
執行編輯　吳孟蓉
編　　輯　許雅眉、程郁庭
特約翻譯　陳姿君
美術主編　潘大智
特約美編　菩薩蠻數位文化有限公司
行銷企劃　謝儀方
出 版 者　橘子文化事業有限公司

總 代 理　三友圖書有限公司
地　　址　106 台北市安和路 2 段 213 號 4 樓
電　　話　(02) 2377-4155
傳　　真　(02) 2377-4355
E － mail　service@sanyau.com.tw
郵政劃撥　05844889 三友圖書有限公司

總 經 銷　大和書報圖書股份有限公司
地　　址　新北市新莊區五工五路 2 號
電　　話　(02) 8990-2588
傳　　真　(02) 2299-7900

初　　版　2014 年 01 月
定　　價　新臺幣 385 元
ＩＳＢＮ　978-986-6062-73-5（平裝）

國家圖書館出版品預行編目 (CIP) 資料

異國風主食料理：焗烤、燉飯、粥品、鍋物等 60 道美味幸福上桌 / 洪白陽作. -- 初版. -- 臺北市：橘子文化, 2014.01　面；　公分
ISBN 978-986-6062-73-5(平裝)

1. 食譜

427.1　　102026809

版權所有 · 翻印必究
書若有破損缺頁 請寄回本社更換

百貨專櫃據點

台北:
士林旗艦店 1F
新光三越台北南西店 7F
太平洋SOGO百貨復興店 8F
太平洋SOGO百貨忠孝店 8F
統一阪急百貨台北店 6F
新光三越台北信義新天地A8 7F
板橋大遠百Mega City 7F
HOLA特力和樂 士林店 B1
HOLA特力和樂 內湖店 1F
HOLA特力和樂 中和店 1F
HOLA特力和樂 土城店 3F

桃園:
FE21'遠東百貨 桃園店 10F
新光三越桃園大有店 B1
太平洋SOGO百貨中壢元化館 7F
HOLA特力和樂 南崁店 1F
新竹:
新竹大遠百 5F
太平洋SOGO百貨新竹店 9F
太平洋崇光百貨巨城店 6F
台中:
新光三越台中中港店 8F
HOLA特力和樂 中港店 1F
HOLA特力和樂 北屯店 1F
台中大遠百Top City 9F

台南:
新光三越台南西門店 B1
HOLA特力和樂 仁德店 2F
嘉義:
HOLA特力和樂 嘉義店 1F
高雄:
新光三越高雄左營店 9F
統一阪急百貨高雄店 5F
HOLA特力和樂 左營店 1F

KUHN
RIKON
SWITZERLAND
瑞康屋
SWISS
MADE
BY KUHN RIKON
DESIGN
PLUS
GOOD
DESIGN
iF
product
design award
2012
瑞康國際企業股份有限公司 TEL 0800 39 3399 FAX 02 8811 2518 www.rakenhouse.com

地址：　　　　縣/市　　　　　鄉/鎮/市/區　　　　　路/街

段　　巷　　弄　　號　　樓

廣　告　回　函
台北郵局登記證
台北廣字第2780號

三友圖書有限公司 收

SANYAU PUBLISHING CO., LTD.

106　台北市安和路2段213號4樓

SANYAU

三友圖書 / 讀者特惠區

購買《異國風主食料理：焗烤、燉飯、粥品、鍋物等60道美味幸福上桌》的讀者有福啦！只要詳細填寫背面問卷，並寄回本公司，即有機會獲得瑞康國際企業股份有限公司獨家贊助之特別好禮！

HOT PAN休閒鍋乙台
（共一名）市價：$9,900
（圖片僅供參考，顏色隨機）

神奇節能板乙個
(共一名)市價：$2,400

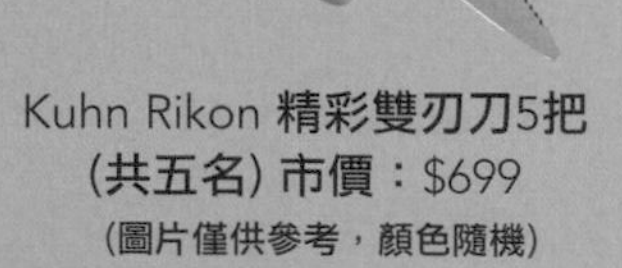

Kuhn Rikon 精彩雙刃刀5把
(共五名) 市價：$699
(圖片僅供參考，顏色隨機)

活動期限至2014年4月30日為止，詳情請見問卷內容。

旗林文化╳橘子文化╳四塊玉文化
www.ju-zi.com.tw
www.facebook.com/comehome

(本回函影印無效)

【黏貼處】對折後寄回，謝謝。

親愛的讀者：

感謝您購買《國風主食料理：焗烤、燉飯、粥品、鍋物等60道美味幸福上桌》一書，為回饋您對本書的支持與愛護，只要您填妥此回函，並於2014年4月30日前寄回本社（以郵戳為憑），即有機會抽中瑞康國際企業股份有限公司提供的精美鍋具與廚具。本活動將於2014年5月7日抽出幸運得主七名。

1 您從何處購得本書？
□博客來網路書店 □金石堂網路書店 □誠品網路書店 □其他網路書店
□實體書店______

2 您從何處得知本書？
□廣播媒體 □臉書 □朋友推薦 □博客來網路書店 □金石堂網路書店
□誠品網路書店 □其他網路書店______□實體書店______

3 您購買本書的因素有哪些？(可複選)
□作者 □內容 □圖片 □版面編排 □其他______

4 您覺得本書的封面設計如何？
□非常滿意 □滿意 □普通 □很差 □其他______

5 非常感謝您購買此書，您還對哪些主題有興趣？(可複選)
□中西食譜 □點心烘焙 □飲品類 □瘦身美容 □手作DIY
□養生保健 □兩性關係 □心靈療癒 □小說 □其他______

6 您最常選擇購書的通路是以下哪一個？
□誠品實體書店 □金石堂實體書店 □博客來網路書店 □誠品網路書店
□金石堂網路書店 □PC HOME網路書店 □Costco
□其他網路書店______ □其他實體書店______

7 您是否有閱讀電子書的習慣？
□有，已習慣看電子書 □偶爾會看 □沒有，不習慣看電子書
□其他______

8 您認為本書尚需改進之處？以及對我們的意見？
____

9 日後若有優惠訊息，您希望我們以何種方式通知您？
□電話 □E-mail □簡訊 □書面宣傳寄送至貴府 □其他______

10 是否願意將個資提供給瑞康國際企業股份有限公司，做為鍋具後續客服使用？ □ 願意 □不願意 □其他

謝謝您的填寫，
您寶貴的建議是我們進步的動力！

（請務必填寫正確資訊，以利獲獎時通知聯繫）

姓名______________ 出生年月日______________
電話______________ E-mail______________
通訊地址______________________________

【黏貼處】對折後寄回，謝謝。